C.R. Tumblin
CONSTRUCTION COST ESTIMATES

Harvey V. Debo and Leo Diamant
CONSTRUCTION SUPERINTENDENTS JOB GUIDE

Oktay Ural, Editor
CONSTRUCTION OF LOWER-COST HOUSING

Robert M. Koerner and Joseph P. Welsh
CONSTRUCTION AND GEOTECHNICAL ENGINEERING USING SYNTHETIC FABRICS

J. Patrick Powers
CONSTRUCTION DEWATERING: A GUIDE TO THEORY AND PRACTICE

Harold J. Rosen
CONSTRUCTION SPECIFICATIONS WRITING: PRINCIPLES AND PROCEDURES, Second Edition

Walter Podolny, Jr., and Jean M. Müller
CONSTRUCTION AND DESIGN OF PRESTRESSED CONCRETE SEGMENTAL BRIDGES

Ben C. Gerwick, Jr. and John C. Woolery
CONSTRUCTION AND ENGINEERING MARKETING FOR MAJOR PROJECT SERVICES

James E. Clyde
CONSTRUCTION INSPECTION: A FIELD GUIDE TO PRACTICE, Second Edition

Julian R. Panek and John Philip Cook
CONSTRUCTION SEALANTS AND ADHESIVES, Second Edition

Courtland A. Collier and Don A. Halperin
CONSTRUCTION FUNDING: WHERE THE MONEY COMES FROM, Second Edition

James B. Fullman
CONSTRUCTION SAFETY, SECURITY, AND LOSS PREVENTION

Construction Safety,
Security, and Loss Prevention

Construction Safety, Security, and Loss Prevention

James B. Fullman, P.E.

A Wiley-Interscience Publication

John Wiley & Sons
New York • Chichester • Brisbane • Toronto • Singapore

This book is based on the author's personal experiences and other sources he has considered reliable, but neither he nor the publisher can make any guarantee of results nor assume any liability in connection with either the accuracy of the information contained herein or the suitability of the suggestions made. Moreover, space restrictions have made it impossible to include many valuable safety procedures. Readers should check important regulations applicable to their work and recognize that changes are constantly being made. No one should feel obliged to agree with the author's opinions if his own experience or evaluation of job conditions leads him to different conclusions. Diversity of opinion is desirable; it is essential to progress.

The author uses the terms "man," "he," "his," and similar words in their generic sense and in every case both the feminine and masculine equivalents are implied.

Library of Congress Cataloging in Publication Data:

Fullman, James B.
Construction safety, security, and loss prevention.

(Wiley series of practical construction guides)
"A Wiley-Interscience publication."
Includes index.
1. Building—safety measures. I. Title. II. Series.

TH443.F84 1984 690′.22 84-5077
ISBN 0471-86821-3
Printed in the United States of America
10 9 8 7 6 5 4 3 2 1

Series Preface

The Wiley Series of Practical Construction Guides provides the working constructor with up-to-date information that can help to increase the job profit margin. These guidebooks, which are scaled mainly for practice, but include the necessary theory and design, should aid a construction contractor in approaching work problems with more knowledgeable confidence. The guides should be useful also to engineers, architects, planners, specification writers, project managers, superintendents, materials and equipment manufacturers and, the source of all these callings, instructors and their students.

Construction in the United States alone will reach $250 billion a year in the early 1980s. In all nations, the business of building will continue to grow at a phenomenal rate, because the population proliferation demands new living, working, and recreational facilities. This construction will have to be more substantial, thus demanding a more professional performance from the contractor. Before science and technology had seriously affected the ideas, job plans, financing, and erection of structures, most contractors developed their know-how by field trial-and-error. Wheels, small and large, were constantly being reinvented in all sectors, because there was no interchange of knowledge. The current complexity of construction, even in more rural areas, has revealed a clear need for more proficient, professional methods and tools in both practice and learning.

Because construction is highly competitive, some practical knowledge is necessarily proprietary. But most practical day-to-day problems are common to the whole construction industry. These are the subjects for the Wiley Practical Construction Guides.

M.D. Morris, P.E.

Preface

What causes accidents? Why does a craftsman have an accident on a certain day while doing his usual job in the customary way? Why do some contractors have fewer losses than their competitors? Are they just lucky? Is it worth the trouble to try to minimize accidents?

Yes, it *is* worth the trouble, and the cost as well. Accident costs have been averaging about 6.5% of construction costs for industrial and commercial projects, and many companies report that their safety programs are highly cost effective. These companies find that safety can increase profits and make them more competitive.

However, hit or miss safety activities can be a waste of money. I am interested in the most efficient as well as the most effective ways of avoiding accidents. There are no pat answers, but there is great room for improvement. Construction has the worst safety record of any industry in the United States.

Many volumes this size would be required to record the usual causes of accidents and give rules for avoiding them. But even if it were possible to include a much more extensive list of specific instructions, this would not be effective. Underlying causes such as the state of mind of the worker that results from the actions of his supervisor or off-the-job influences must be understood and considered.

This is a guide book, not a book of rules. I believe it is better to illustrate the ways in which people have been successful in solving their problems than to try to provide all the answers. Some parts of the book deal with matters that affect safety but cannot be readily reduced to rules. If accidents are to be minimized, building owners, contractors, superintendents, foremen, and inspectors need to be aware of how their attitudes and actions affect safety. To discuss such matters, I have had to depart at times from the usual approach to accident prevention. If by so doing I have stimulated thoughts about the basic causes of accidents, these apparent digressions will have been justified.

Safety is a broad subject; it needs to be broken down into manageable parts and considered at the appropriate times. Therefore, I have arranged topics in the approximate sequence in which it would be reasonable to consider them during the construction of a typical building. The reader who has the opportunity to take formal courses in industrial hygiene, insurance, structural safety, fire prevention, and similar subjects will find them valuable, but I assume that most readers will have to dig out answers for themselves. I have indicated sources of information that should be useful. Anyone who gets satisfaction from making a valuable contribution to the financial success of his organization and at the same time from helping workers to avoid the pain and losses resulting from accidents can learn to plan and execute an effective safety program.

No book about safety can ignore government regulations, yet nobody could include them all in a book of reasonable size. In some instances I have summarized highlights and then given a specific reference to sources. A few regulations are in the appendices. If regulations apply to the performance of certain hazardous jobs that are usually done only by specialists, I have assumed that those individuals were aware of the rules, making it unnecessary to include them in a general guidebook on construction safety.

I have commented on some security and loss prevention problems because I consider them closely related to safety. Some readers may consider these topics to be digressions, but I have observed that they usually warrant attention as part of an effective safety program.

JAMES B. FULLMAN

Stamford, Connecticut
July 1984

Acknowledgments

I am indebted to many people for their comments and constructive criticism. It is inevitable that I will have inadvertently neglected to mention some whose help should be formally recognized, and to them I extend my apologies.

Everett Smethurst, Editor, Wiley Interscience Division and M.D. Morris, P.E., Series Editor, have been most helpful and encouraging.

Special thanks are also due to H.M. Steinmetz, attorney, of Riverside, Connecticut; George Yousko, Construction Safety Advisor, Exxon Research and Engineering Co.; A.C. McLean, Construction Engineer, Philadelphia Electric Co.; Andres van der Poll of McDonald and van der Poll, consulting structural engineers; Gerald Kolinsky, Connecticut Workmen's Compensation Commissioner, 7th District; J.D. Holthusen, Corporate Manager Fire Protection, J.C. Penney Co.; D.F. Mahoney of Taylor-Lawlor Insurance Associates, Inc.; Stewart Campbell of Her Majesty's Health and Safety Executive in London; L.R. Deal, Regional Loss Control Services Manager, Wausau Insurance Companies; Dr. Leo Donovan, Director, Du Bois Day Treatment Center, Stamford, Connecticut; Rev. E.J. Hussey, Dioces of Bridgeport, Connecticut; Dr. J. Gross, Podiatrist, Stamford, Connecticut; Thom Serrani, State Senator of the Connecticut General Assembly; R.M. Kaletsky, Supervisory Safety Specialist, OSHA, Hartford, Connecticut; J.C. Kaussner, Copper Development Association; J.V. Hansen, Member of Congress, Utah; S.M. Barone, National Council on Alcoholism, New York City; G.M. Wellington, New York Regional Director, National Safety Council; R.L. Voss, Lieutenant Commander, USNR, Retired; and to W.F. Chen, X.L. Liu, S. Karshenas, A.H.S. Ang, and A.P. Arndt, all of whom kindly supplied copies of papers presented at the 1981 ASME convention.

I also want to acknowledge my debt to friends and former associates at Olin Corporation, particularly the late Sidney Sussman, Ph D; Wm. J. Maher; Sue Ellen Klein, and Thomas Smith.

Sources of photographs of safety equipment are indicated in the captions, but I want to recognize also the personal help of M.J. Loftus of Klein Tools; J.C. Nelms of Miller Equipment; D.E. Maurer of Willson Safety Products; D.K. Andrew, Jr. of L.L. Bean Co.; G. Mohan of International Shoe Co.; J.J. Kritek of Skil Corp.; G.W. Bersett of Ramset Division, Olin Corp.; and F.C. Ferguson of Mine Safety Appliances Co.

The final review suggestions by Ralph W. Niles, Safety Engineer, Volpe

Construction Co.; Thomas W. Floyd, Federal Occupational Safety and Health Specialist; Thomas M. Mihalik, Federal Safety Engineer, and by my patient and understanding wife, Helen-Louise, were all of inestimable value. I greatly appreciate their efforts.

In recognizing the assistance of others, I do not wish to imply that those mentioned necessarily agree with all the statements I have made. Further constructive criticism will be welcomed.

J.B.F.

Contents

1 A Quick Look at Construction Industry Safety 1

1.1 *How Safe is Construction Work? 1*
1.2 *Why Does Construction Have a Poor Safety Record? Must it Continue? 3*
1.3 *How Much Do Accidents Cost? Who Pays for Them? 6*
1.4 *What Causes Accidents? 8*
1.5 *Can Construction Work Be Safe? 13*
1.6 *Do Safety Regulations Reduce Accidents? 15*

2 Planning a Safety Program 18

2.1 *General Strategy 18*
2.2 *Motivation 21*
2.3 *Knowledge and Awareness 31*
2.4 *Monitoring and Adaptation 47*

3 Preconstruction Preparations 50

3.1 *Applicable Regulations and Responsibility for Enforcement 50*
3.2 *Insurance 53*
3.3 *Anticipating Emergencies 66*
3.4 *Community Relations 79*
3.5 *Personal Safety Equipment 82*
3.6 *Training Activities and Problems 94*

4 Site Preparation 98

4.1 *Site Inspection 98*
4.2 *Protection of the Public: Restricting Access to the Site 99*
4.3 *Demolition Work 100*
4.4 *Use of Explosives 103*
4.5 *Clearing Work 104*
4.6 *Trash Disposal 106*
4.7 *Underground Obstructions 108*
4.8 *Surveying and Staking Out 108*

5 **Excavations, Utilities, and Foundations 110**

5.1 On-site traffic 112
5.2 Earth-moving Equipment 113
5.3 Handling and Driving Piles 116
5.4 Compressed Air 117
5.5 Shoring and Sloping of Trenches and Excavations 118
5.6 Laying Pipe in Trenches 122
5.7 Water Supplies 123
5.8 Electric Power 127
5.9 Other Utilities 130
5.10 Placing Concrete Footings and Foundation Walls 130

6 **Erecting and Closing In the Structure 134**

6.1 Safety Program Review and Revisions 134
6.2 Materials Handling 136
6.3 Falls 155
6.4 Fire Damage Prevention and Control 171
6.5 Roofing 179
6.6 Avoiding Structural Collapse 183
6.7 Glass and Facade Failures 193

7 **Installing Mechanical Systems and Completing the Interior 194**

7.1 Labor Conflicts 194
7.2 Positioning Parts of Systems 196
7.3 Small Tool Safety 197
7.4 Cutting and Patching 206
7.5 Housekeeping 207
7.6 Security 209
7.7 Confined or Enclosed Space Hazards 211
7.8 Testing Mechanical and Electrical Systems 219
7.9 Painting 224
7.10 Inspections 227

8 **Finishing the Job 230**

8.1 "Changing of the Guard" 230
8.2 Inspection and Acceptance Reports 232
8.3 Insurance and Security 233
8.4 Service Agreements 234
8.5 Safety Awards, Ceremonies, and Publicity 237

***Appendices* 239**

A. *OSHA Recordkeeping and Incidence Rates* 239
B. *OSHA and Bureau of Labor Statistics Regional Offices* 244
C. *Some Dangerous Substances* 246
D. *Shade Numbers for Welder's Goggles or Face Shields* 252
E. *Electric Extension Cord Capacities* 253
F. *Illumination Required in Work Areas* 254
G. *Respirator Selection for Various Hazards* 255
H. *Scaffolds* 258
I. *Job-Built Cleat Type Ladders* 261
J. *Sample of Confined Space Entry Permit* 264
K. *Checklist of Construction Safety Hazards* 266
L. *Author Biography* 279

***Index* 281**

Construction Safety,
Security, and Loss Prevention

1

A Quick Look at Construction Industry Safety

1.1 *How Safe Is Construction Work?*

One in every six construction workers suffers an occupational injury or illness on the average of once a year. One in 16 has a lost time injury, and about one in 3300 has a fatal accident or job-related illness. Construction work and mining compete for the worst safety record. Miners have more fatalities per thousand workers; construction workers have more injuries. The average construction worker loses about 1.2 days per year as a result of injuries, and in 1979 almost 1000 workers lost their lives.

Transportation and public utility workers have one third fewer accidents than those in construction and less than 40% as many fatalities per thousand workers, but news of a plane crash spreads immediately because many people are involved at the one location. Construction accidents, with a few notable exceptions, tend to pick off workers one at a time. The local newspaper carries a paragraph or two about the accident, but that is about all anyone ever hears about it. Imagine the public furor if 1000 workers were killed in a nuclear power plant accident, or if there were many such accidents involving a few workers at each site. Public outcry would be deafening and every town would be up in arms. Yet this is exactly what happens in the construction industry, and nobody screams.

The U.S. Department of Labor, Bureau of Labor Statistics Construction Industry Report is the source of data used to prepare Figures 1.1 to 1.3. Figure 1.1 compares the annual number of injuries per 100 workers in several industries. The average incidence rate for construction was 16.2 per 100 workers, indicating that about one out of every six construction workers was hurt during the year. No other major industry has more accidents per 100 employees.

Construction	XXXXXXXXXXXXXXXX	16.2
Manufacturing	XXXXXXXXXXXXX	13.3
Agriculture, Forestry, and Fishing	XXXXXXXXXXXX	11.7
Mining	XXXXXXXXXXX	11.4
Transportation and Public Utilities	XXXXXXXXXX	10.1
Wholesale and Retail Trades	XXXXXXXX	8.0
Services	XXXXXX	5.5
Finance, Insurance, and Real Estate	XX	2.1

Source: Bureau of Labor Statistics Annual Construction Industry Report, April 1981; (1978–1979 data).

Figure 1.1 Incidence rates, injuries, and illnesses for various industries (number of cases per year per 100 full-time workers).

Figure 1.2 shows that this cannot be blamed on any single sector of the industry or trade. No sector has a good record; some are just worse than others. Even mining, with an incidence rate of 11.4, looks good by comparison.

Figure 1.3 compares fatalities in different industries. Construction still does not look good. Mine disasters often involve large groups of workers, and I assume that is the reason the industry's rate is higher than for construction in spite of its lower incidence rating. Note that construction workers were six times more likely to be killed on the job than workers in manufacturing.

It is often claimed that the nature of construction work makes accidents inevitable. Don't believe it; dig beneath the platitudes and you will find that some employers know how to avoid accidents; their safety records prove it is possible to greatly improve the industry average. I think many safety programs have been ineffective because they were based on faulty assumptions about the causes of accidents and, as a consequence, the wrong cures were proposed. In my opinion too much attention has been paid to the mechanics of accident prevention, and too little to the intangible personal influences. In the following chapters I will discuss both approaches to the problem of preventing accidents.

Construction Industry Average	XXXXXXXXXXXXXXXX	16.2
General Building Construction	XXXXXXXXXXXXXXXX	16.3
Residential Building Construction	XXXXXXXXXXXXX	13.0
Heavy Construction Contractors	XXXXXXXXXXXXXXXXX	16.6
Plumbing, Heating, and Air Conditioning	XXXXXXXXXXXXXXXXX	17.0
Painting, Paperhanging, and Decorating	XXXXXXXXXX	9.6
Electrical Work	XXXXXXXXXXXXXX	14.0
Masonry, Stonework, and Plastering	XXXXXXXXXXXXXXXX	16.4
Carpentering and Flooring	XXXXXXXXXXXXXX	14.0
Roofing and Sheet Metal	XXXXXXXXXXXXXXXXXXXXXXX	22.3
Concrete Work	XXXXXXXXXXXXXXX	13.8
Water Well Drilling	XXXXXXXXXXXXXXX	15.2

Source: Bureau of Labor Statistics Annual Construction Industry Report, April, 1981 (1979 data).

Figure 1.2 Incidence rates for construction, by types of construction work and trades. Number of injuries and illnesses per year per 100 full-time workers.

1.2 Why Does Construction Have a Poor Safety Record? Must It Continue?

A lot of money is wasted on safety programs that contractors don't expect to be effective and that workers ignore. Let's stop kidding ourselves. If contractors who think accidents are unavoidable would tell their employees to look out for their own safety, everybody would know exactly where he stood. That approach would also make it easy to identify contractors who felt that a real safety program would contribute to profitability. Worker cooperation would be easier to obtain. It is the present prevalence of half-baked plans that creates a cynical attitude toward safety. When people think nothing can be accomplished, they are probably right.

Americans are fortunate in being able to have just about anything they want if they want it badly enough. If workers, the general public, and

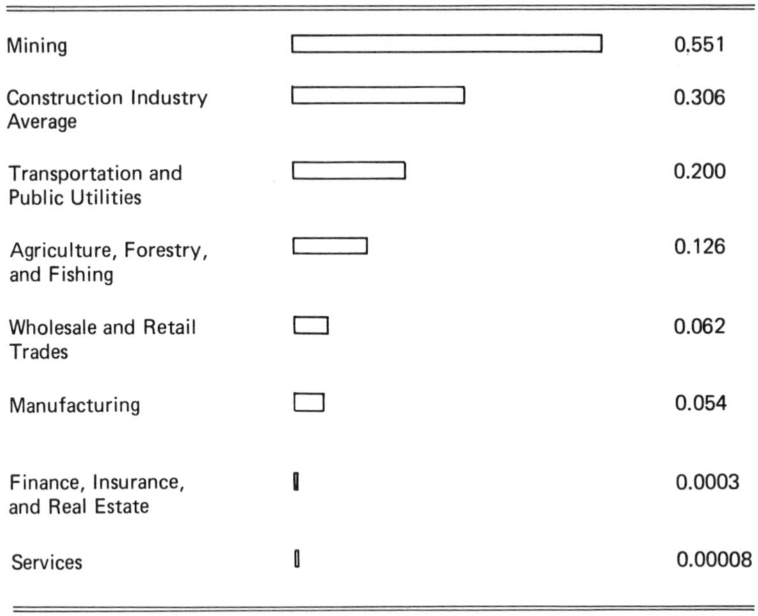

Source: Calculated from data in Annual Construction Ind. Report, April 1981, Bureau of Labor Statistics.

Figure 1.3 Number of fatalities per year per 1000 workers in various industries.

contractors want a lower accident rate, they can have it. There is no doubt about it; many contractors have demonstrated that a very high percentage of accidents are preventable. Everybody recognizes that there are special safety problems in the construction industry, but people can solve problems when they put their minds to it. It is just a matter of priorities. Contractors and workers already have access to knowledge about how to avoid most accidents, but if construction workers themselves rank safety below wages and fringe benefits is it any wonder that contractors, public officials, and even union leaders blow hot and cold on the subject?

Safety hasn't been sold well enough. It can't be carried home like groceries in a bag, but like the grocery bag, we only give it a thought when it breaks. We only appreciate health when we lose it. People do want safety, but they just forget to keep on wanting it enough. Only a few people make their livings selling safety. Insurance companies want to reduce accidents, but they would be out of business if they were successful in eliminating them. A few industrial hygienists and safety engineers can't do an effective job of promoting safety without the active support of the public, employers, and government.

The regulatory route to safer construction work has been disappointing. OSHA would need a full-time compliance officer on every construction job to enforce all the regulations that have been issued. Except in a police state, it is impossible to enforce laws without the consistent support of the public and the people directly involved. Consider the problem of drunken driving; in the United States it is almost, but not quite, acceptable behavior. People drive when their licenses have been suspended and they are seldom severely punished for it. Friends condone their actions, and acquaintances do not think they have done anything very wrong even when they are involved in an accident. We will keep on allowing 25,000 people to be killed on the highways every year by drunken drivers until public opinion demands that it stop. In Sweden, it has been stopped. Swedes still drink as much as ever, except for the one driver per car who stays sober. No system ever works perfectly, but it works remarkably well there because people want it to work. Offenders spend enough time in jail to deter them from becoming repeaters.

With everybody's support, OSHA would probably be effective in sharply reducing construction accidents, but the administration has little backing and industry spends more time looking for loopholes in the law than in trying to comply with the intent of the regulations. Construction is a business, not a crusade. Unless safety is perceived as a means of increasing profits, who is going to get excited about it? Some contractors have found that safety pays, but it is one of the best-kept secrets of our time. Congress and the courts have made a mistake in ruling that safety regulations need not be cost effective. Bureaucrats can ignore economics, but they cannot keep others from looking at cost/benefit ratios. If it is more profitable to pay fines and other costs of accidents than to follow all the regulations, what can be expected?

Of course there are a few idealists who expect ethical behavior to prevail and who believe that everybody should avoid accidents because it is the humane thing to do. They are the salt of the earth, but they cannot survive in the day-to-day scramble to make a living in the construction industry unless they have learned how to get more than a dollar's worth of savings for every dollar spent to promote safety. Some of the largest and most profitable construction companies are experts in safety. They recognize that in order to get jobs done efficiently they must minimize accidents.

Cost effectiveness in safety is difficult to achieve where plan examiners and inspectors are willing to overlook safety hazards in return for gratuities or other favors. It is no secret that in many, but by no means all, communities such practises are considered normal. Who would expect fewer accidents if it is more profitable to circumvent safety regulations than to concentrate on avoiding injuries?

I am convinced that safety pays if the costs are correctly audited, and that there will be many fewer accidents if those who fail to prevent them must pay the full cost of all avoidable incidents. This does not seem to be the case at the present time, but pressure from voters, unions, insurance companies, legislators, and courts can change the ways costs are shared by contractors, workers, and the general public.

The worker's compensation system needs to be improved, and changes are being made on a piecemeal basis, state by state. Liability laws can certainly be improved. Many regulations could be made more cost effective without necessarily adding to hazards. As soon as we concentrate on doing what is most efficient, we can expect a marked improvement in construction safety. But, first, let us consider what accidents actually cost.

1.3 How Much Do Accidents Cost? Who Pays for Them?

The Stanford University Department of Civil Engineering, under contract to the Business Roundtable, has done extensive research and analysis of the cost of accidents. Their findings are in that department's Technical Report #260, "Improving Construction Safety Performance: The User's Role." In 1979, accidents cost about $8.6 billion, or about 6.5% of the $137 billion spent for industrial, utility, and commercial construction. The researchers concluded that it would be reasonable to expect that good safety programs of the type being used by certain construction firms could reduce annual accident costs by about $2.75 billion. The probable cost of such programs would be about $0.86 billion, making them cost effective by a ratio of 3.2 to 1.

Worker's compensation insurance premiums are the most easily recognized measure of the cost of accidents. Premiums frequently range between 5 and 50% of straight time wages; in Washington, D.C., for example, the rates for wreckers and pile drivers exceed 75% of wages. Any contractor who looks at the amount he has to pay for this item must recognize that accidents cost the industry dearly. In Section 3.2 where insurance costs are discussed in more detail, I have pointed out that insurance companies are usually willing to give substantial discounts to contractors who have good safety records. These discounts are often more than enough to pay for good safety programs, even when it is recognized that they never equal the total savings to the insurer.

Following an accident, compensation insurance and medical payments to the injured employee are obvious costs that must eventually be paid by the employer if he stays in business long enough. No insurance company will continue to carry an account that is unprofitable. But investigators at Stanford University's Department of Civil Engineering have proved that total accident costs are apt to be 4 to 17 times direct costs, including insurance. The benefits received by the injured worker from the insurance carrier represent only about 65% of insurance cost. Indirect costs of accidents include:

Investigation expense. Preparing reports takes the time of supervisors who would otherwise be doing more productive work. In addition, consultants may have to be retained.

Fines may have to be paid, or citations may have to be protested in the courts.

The job may be delayed. There could be penalties for failure to meet completion dates, and money can be tied up that could be better used elsewhere.

There may be damage to the structure or waste of material. Even if covered by insurance, processing of claims is an expense and most policies do not cover total losses.

Overtime may have to be paid to other workers.

A replacement for a disabled worker may have to be hired and trained. Meanwhile, productivity decreases.

Any contractor who allows accidents to increase his costs puts himself at a competitive disadvantage, because he has to bid for more jobs to get the same amount of business, and he may end up doing work that other contractors do not consider attractive.

When an accident may have been caused by the negligence of anyone other than the worker's employer, the worker may sue the responsible party. Even though that possibility has been anticipated and insurance has been purchased, it may not be sufficient to cover awards made by sympathetic jurors plus the cost of legal fees and other costs of litigation. Adequate liability insurance is expensive; it adds to the cost of every structure.

In addition to the foregoing costs to contractors, design professionals, and owners, injured workers bear many of the costs of accidents. Worker's compensation insurance benefits are substantially less than regular wages; therefore it is only occasionally that an injured worker who sees no probability of finding work after he recovers can profit by stretching out his recuperation period and continue to collect benefits to which he would not normally be entitled. In more typical cases, injury results in a financial loss to the worker. I have no way of knowing what percentage of normal income is likely to be lost, because many workers have part-time jobs that they lose as a result of being unable to do their regular work, and extra family income may be lost if a spouse must lose days from work. If disability is permanent, the opportunity to start a business or advance to supervisory work will generally be lost.

Some readers may recall large awards to workers as a result of legal action against third parties, and it is sometimes popularly assumed that these are typical. They are newsworthy because they are unusual; moreover, the present system of compensation is inequitable: it allows one worker to be generously compensated for an injury and fails to deal similarly with another who suffered a comparable injury but who could not find a basis for legal action against a third party. The uncertainties inflate the liability insurance premiums and are a cost of accidents that may be paid by either or all design professionals, contractors, and owners.

Workers expect to be paid for taking risks. They may not say much, but when work is considered dangerous, qualified people can demand higher wages. Why should a skilled mechanic risk injury unless times are so bad that he is desperate for work or the pay is too tempting to refuse? Although there are usually more applicants than jobs, their qualifications are not necessarily impressive. I cannot prove that low productivity, high wages, and dangerous working conditions are related by a mathematical formula, but they seem to be strongly interdependent.

If one takes a broad overview of the cost of accidents, it is impossible to avoid the conclusion that they are a drag on the industry. Roads and bridges are badly in need of repair. People need housing. Industries need new buildings. There is plenty of work to be done at an affordable price, but that requirement is not being met in many instances. Reducing accidents can lower costs, and every step in that direction is worth taking. Nobody profits from contracts that are never awarded.

1.4 What Causes Accidents?

Figure 1.4 shows the kinds of accidents that are most likely to cause fatalities in the construction industry. The typical accident can be readily assigned to one of these categories, but that is just the first step in determining its cause.

In order to avoid similar occurrences in the future much more detailed information is needed, though it may be difficult or even impossible to obtain. Witnesses may not want to become involved because they fear they may have to spend a great deal of time in court. Employers and supervisors may consider it unwise to release information without legal advice. People may not even agree about what happened and in such cases agreement on the cause of the accident is very unlikely. Fuller and more timely dissemination of information about the causes of accidents would make it easier to prevent them.

Reports should include more details about the underlying causes of accidents. For example, assume that a worker falls from a ladder. The accident would be readily assigned to the "falls" category, but we would still know little about its cause. Did a rung break? Did the ladder shift? Was the worker carrying tools in his hands when he fell? Did he have a dizzy spell? Was he taking medication? Was he well enough to be at work?

An accident may have several causes. In the foregoing example the direct cause may have been shifting of the ladder or failure of the worker to be sure it was secure. Another obvious cause may have been illness that made him dizzy and less attentive to safety precautions than he should have been. An indirect cause may have been the use of a ladder for a job that should have been done from a scaffold. A contributing cause may have been failure of his supervisor or fellow workers to observe that he was not well enough to be at work or that the ladder was insecure.

Falls	XXXXXXXXXXXXXXXXXXXXXXXXXXXXX	29% *
Over-the-road Motor Vehicles	XXXXXXXXXXXXXXXX	16%
Industrial Vehicles and Equipment	XXXXXXXXXXXXX	13%
Electrocution	XXXXXXXXXX	10%
Caught In, Under, or Between Objects	XXXXXXXXX	9%
Heart Attacks	XXXXXXXX	8%
Struck by Objects	XXXXXX	6%
Explosions	XXX	3%
Fires	XX	2%

* Total is less than 100% becauses individual causes responsible for less than 1% of fatalities were not included

Source: U.S.Dept. of Labor, Bureau of Labor Statistics, Report number 586, March, 1980.

Figure 1.4 Principal causes of construction industry fatalities.

In classifying causes, I see no great need to distinguish rigorously between direct, indirect, and contributing causes of accidents, but I do stress the importance of searching for all of the causes. Be sure to recognize that an accident might be the result of two concurrent events, neither of which would necessarily be dangerous in itself. For example, a worker may step on a hose, and another worker may pull on the same hose. If both events happen at the same time, one of the individuals may suffer a serious fall.

Some analysts differentiate between mechanical and human failures as causes of accidents. Material defects and equipment failure are thus considered mechanical causes, and errors of judgment or illness human causes. But is it a mechanical or a human failure when an inspector fails to observe something that is wrong with a machine?

This question may often be asked when someone is "struck by a falling object." For statistical purposes this classification is useful, but it does not help much in preventing accidents unless the investigation discloses (1) what fell, (2) why it fell, and (3) why someone was in a spot where he could be struck by it.

Consider a recent "struck by" fatality. A trucker and his wife had brought a load of steel to a job. While a crane was unloading the material the woman waited in the cab of the truck. The crane boom collapsed, dropping over 10 tons of steel onto her. The load should have been divided into three lifts instead of the two lifts which overloaded the crane. Charges of negligent criminal homicide were pending for a year against two ironworkers, but were dropped for lack of evidence. I have not seen an explanation of why the crane operator failed to detect the overload. All causes of an accident must of course be investigated, but I think the most important question in this case is why anyone was permitted to sit in the cab during the unloading. The ironworkers, crane operator, driver, and even the deceased herself should have had enough common sense to recognize the dangers. If the cab had been empty, only a simple claim for damage to the truck would have been involved.

Some people consider accidents inevitable and think prevention is futile. This viewpoint is of course diametrically opposed to the idea that every accident has one or more causes, and that by eliminating the causes, all accidents can be prevented.

Regulatory bodies tend to follow the cause-effect line of reasoning, with primary emphasis on the mechanical as opposed to the human causes of accidents. Consequently, regulations usually cover dimensions of lumber, allowable stresses, and other matters that can be measured or calculated. When these regulations attempt to reduce the human contribution to accidents by requiring workers to take training courses or the licensing of operators, they are likely to be ineffective. Errors of judgment or momentary physical impairment cannot be written away by rules. Therefore, accidents still occur even when there are no code violations. When a worker falls from a perfectly secure ladder while climbing it in a proper manner, he may be having a minor stroke. As far as I have been able to determine, periodic medical examinations may indicate that a person could be at risk but cannot predict with any degree of certainty that such illness can be expected on a certain day or within a certain time interval.

Imagination is needed by anyone who hopes to anticipate and thereby minimize accidents. The essence of supervision is the ability to see what others overlook, literally to have *super*ior vision. A good supervisor notes little signs that a worker is not up to his usual self and keeps an eye on him. That is not the sort of activity that can be written into a regulation, but it is an important factor in reducing accidents.

Workers often have accidents while doing the same job they have been doing for years in just about the same way. Why, on a certain day, does he get hurt? As so often happens, the worker may have been working in an unsafe manner from the start, leading to an accident that was bound to occur sooner or later. For example, if a worker habitually leans a ladder against a rafter, causing it to rest with one rail against the rafter and the other floating free, it may not fall until after many days, but it has been unsafe from day one.

Illness is often the cause of accidents; minor ailments can be sufficient to impair judgment of distance or cause distractions. Pollen in the air disturbs some people. Their eyes water, and they have all the symptoms of a bad head cold. A simple sneeze at the wrong moment may be enough to cause injury. Safety requires forethought; the worker should recognize that on a certain day he has to take a little more than the usual care while doing certain jobs. Figure 1.4 indicates that heart attacks are responsible for 8% of on-the-job fatalities. Those who have survived heart problems can recall early warning symptoms such as chest pains, muscle cramps, and dizzy spells that could have caused accidents. The same symptoms may be associated with minor and more serious illnesses. It seems to be in human nature to avoid seeing a doctor if we suspect that he may discover an ailment that will disrupt and threaten our lives. We hope that our troubles will disappear if we ignore them. Most of all, we do not want anyone else to know, particularly if his knowing could affect our job security or our freedom to do some things we want very much to continue to do. The modern trend toward having tests made at a hospital makes it harder to conceal one's illness and, consequently, there is more resistance on the part of some workers to see a doctor until it is absolutely necessary.

The temporary discomforts that often follow a night on the town are also clearly common causes of accidents. I refer to the headaches and related effects of overindulgence and not chronic alcoholism. The worker may try to prove he is in perfect shape, and get in trouble fooling himself. (Note that alcoholism is now generally recognized to be a disease. I discuss it in Chapter 2, Section 2.2.)

No athlete would exert himself fully without a little warm-up, but a worker may tackle a hard job the first thing in the day, hoping to get done with it while he is feeling rested. Back injuries, sprains, and similar conditions can be attributed to this cause.

Later in the day, fatigue, heat exhaustion, and sun stroke may cause accidents. Roofers are known to be prone to these stresses, but there is no reason to entirely discount their effect on other craftsmen.

Weather conditions are seldom considered to be under the contractor's control, even though days lost as a result of weather are recognized as a scheduling problem. But in the prevention of accidents, more attention should be given to microclimatology. Wind velocity can be significantly higher at the top of a structure than at ground level. Doubling of the wind speed makes a four-fold increase in the force it exerts on a worker and on anything he may be handling. To avoid accidents, it may be necessary to suspend some operations even though other work can proceed in safety. In some industries, such as air transport, weather forecasting and local observations are matters of great concern; nevertheless, accidents still occur. Contractors cannot ignore the possibility of rapid changes in the weather without significantly increasing the risk of an accident. Wind speed alarms are available, and for a very few dollars a battery-powered weather radio can be purchased. In the metropolitan New York area, such radios are tuned to stations that broadcast weather reports continuously 24 hours of the day. The receivers cannot be tuned to other stations.

Throughout the rest of this book I refer to a number of physical hazards mentioned in various regulations. They may properly be considered causes of accidents, but it would be repetitious to comment on them at this point.

Even after all the apparent causes of accidents have been considered, it is still impossible to explain all accidents from a physical cause and effect relationship. The mental attitude of the worker at the moment is, I think, the missing link that explains why an accident occurs at one particular time but not at other times when conditions are apparently exactly the same.

In the *Statistical Abstract of the United States,* published by the U.S. Department of Commerce, Bureau of the Census, 1981, the death rates for males, 15 to 24 years of age were listed as 100.5 per 100,000 population as a result of accidents, and 20.0 as a result of suicide. For older age groups, up to a retirement age of 65, accident rates declined about 35% but suicides increased about 25%. Mental health problems are a serious threat to present day industry. The suicide rate indicates the extent of extreme depression, but there are many degrees of depression varying from simply feeling a bit "down" to complete breakdown of a person's desire to go on living. Initial signs of trouble might be tardiness, absenteeism, general indifference, or sluggishness. Sometimes the worker may make self-depreciating remarks, commenting, for example, that he is "over the hill" when there is really nothing wrong with him physically. Mental illness can be masked by drugs or alcohol, both of which can be used at the same time; their effects are hard to distinguish from each other.

When a person is depressed, he has lost confidence in himself and has little self-respect. There may be many causes. Even professionals do not always agree on terminology, and I do not want to get involved in clinical analysis of psychological symptoms, but a person does not have to be a doctor to recognize that personal troubles and worries may preoccupy him and make it difficult to concentrate on the task at hand. Although marital difficulties, financial setbacks, fear of layoff, worry about physical problems, and anxiety about the future are all typical problems that just about everyone encounters at one time or another, these problems may become exceptionally hard to face at times or everything may seem to go wrong at once; it is at these times that the ability to work safely can be impaired. Paradoxically, many patients may feel glad that an accident happened to take them out of a situation that they could no longer handle, and there may be a correlation between accidents and the stresses to which people are exposed on or off the job.

The long-term effects of psychological problems can have an important bearing on construction safety. In addition, short-term problems can be the underlying cause of accidents. How many of us have floored the accelerator in anger when someone honked his horn because we were a second late moving away from a red light? We all are apt to overreact at times and, if it is impossible to get at the offender, we may do something foolish. On the job, reaction to a chance remark may cause a worker to throw down a tool in anger or put entirely too much muscle behind any job he may be doing. What passes for kidding may have too much edge behind it, particularly if there is basic animosity between the individuals involved. Control of the way in which workers get along with each other is largely up to their immediate supervisor. He cannot expect everyone to be on the best of terms with everyone else all the time, but he should certainly be aware of serious enmity before it breaks out of control. I recall incidents of knife fights on the job, and a shooting that was narrowly averted by intervention of fellow workers. These incidents could not be considered accidents, but the possibility that the same emotional steam could have caused an accident cannot be overlooked.

1.5 Can Construction Work Be Safe?

Figure 1.5 shows remarkably different accident incidence rates for contracting firms of different sizes. Note that the smallest and the largest firms have the best safety records, which is true of all three industries for which data are plotted. The complete explanation for these observed variations is open to conjecture; I choose to think that the small and large outfits probably care more about safety than the others, but for different reasons.

Companies with only a few workers may consist of members of the same family. If the firm gets all the work it needs to stay in business and has no particularly strong incentive to expand, the same crew may have been doing the same type of work for many years. Roofers, plumbers, and sheet metal firms are often in this category. Over the years, unsafe workers have probably been weeded out while those remaining have improved their skills. In addition the better a worker is at his job, the less apt he is to have an accident.

In a very large organization, the size and complexity of the projects the firm is likely to handle makes it obvious that safety is important. Senior executives have probably had experience with litigation following accidents; they realize that accidents are costly. They also recognize that the assets of a large company make it a target for negligence suits under the slightest pretext.

Large firms can also afford the cost of well planned safety programs and are likely to have learned how to make them effective.

The greatest opportunity for improvement of safety records seems to be in the intermediate sized companies. Perhaps they have too many accidents because they attempt work for which they are inadequately prepared. They may have had to expand by hiring workers and put them on the job before they could be trained to work safely. In many instances,

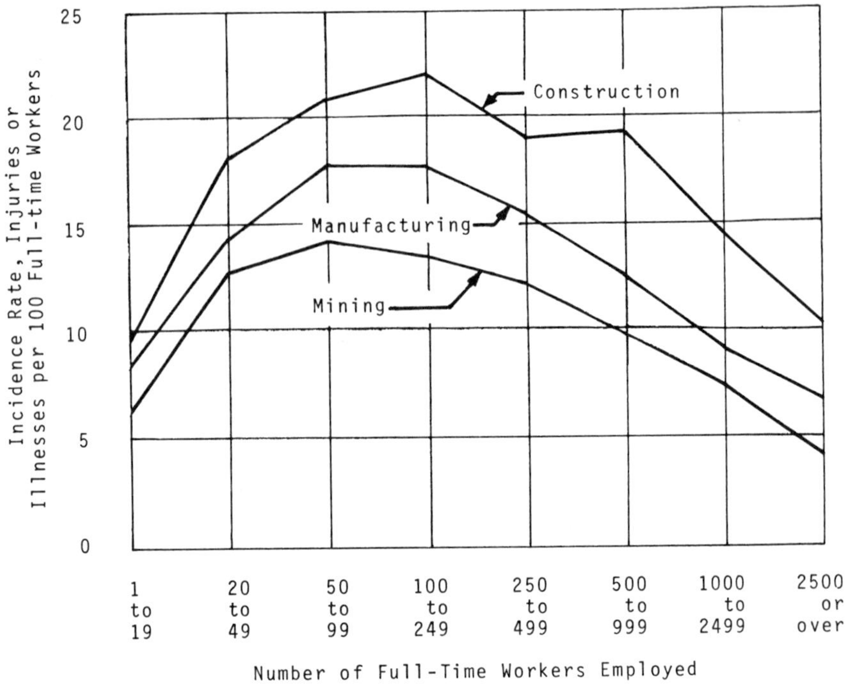

Source: Drawn from data in U.S. Department of Labor Annual Construction Industry Report, April, 1981.

Figure 1.5 Effect of size of company on occupational illness and injury incidence rates for three industries.

nobody in the company may fully realize the importance of safety. It is quite possible that some firms have had to remain smaller than they would have preferred, simply because their poor safety experience and correspondingly high overhead expenses have kept them from being effective competitors with the leaders in their industry.

When some contractors can achieve safety records that are less than one third as high as the average of their competitors, it proves that construction work can be done more safely. These firms had lower incidence rates than the average of either manufacturing or mining, both of which have better opportunities to avoid certain kinds of accidents by engineering means. But it is not easy to avoid accidents. It takes dedication, effort, and persistence. Management skills are needed, just as they are required in improving productivity. Reducing losses is one of the best ways to increase profits; a dollar saved is a dollar earned.

Bechtel, currently the largest contractor in the country, has a safety record well below 50% of the industry average on a $4 billion project for Philadelphia Electric. Exxon Corporation, the world's leading oil company, exercises tight control over the safety programs of contractors and subcontractors on its jobs. On a recent project that took 24 million man hours to complete, lost time accidents and compensation payments were only 10% of the industry average, and there was not a single fatality. Offshore drilling for oil is probably the most hazardous activity in the construction field, with an average fatality rate three times as high as for construction work as a whole. Yet the Glomar Pacific, a 452 foot long ship with a drill rig amidships, worked nearly four years without even one lost time accident or fatality. Exxon safety engineers had a hand in that operation also, although the ship's owner was Global Marine, Inc. More publicity should be given to these records. It is only occasionally that the news media tell the good side of the safety story; it would be a cheering way to start the day instead of going off to work with a cap full of accident and robbery and assault stories to think about.

1.6 Do Safety Regulations Reduce Accidents?

Rules are useless unless they can be understood and enforced. Some of the best rules ever written were the shortest. The 10 commandments were brief enough, but the golden rule covers substantially the same ground in even fewer words. About 4000 years ago, Hammurabi decreed that if a house fell down and killed the owner, the builder should be put to death. He laid it right on the line. Now, even specialists are hard pressed to keep up with regulatory matters.

OSHA's book of regulations applicable to construction contains about 600 pages and, if all the standards incorporated in it by reference were to be counted, the total number of pages would be much greater. Some people's activities have to be regulated; they have no personal integrity and there seems to be no way to keep them in line without enforcement of laws. Therefore, it must be concluded that safety regulations do prevent some accidents.

However, it is debatable whether current safety regulations minimize accidents rather than simply reduce them. I think the construction industry could do with fewer laws if contractors were required to achieve certain results without being burdened with so many instructions about how the results must be achieved.

When regulations become too specific with regard to construction details, they become maximum as well as minimum standards. Who can blame owners for objecting to paying for more than the law demands? If smoke alarms are mandatory and all builders are installing them, their cost will be part of the bid, but if they are not required, why should an owner not leave it to his tenants to install whatever alarms they consider necessary? On the other hand, why should a tenant be aware of hazards that professionals ignore? Similar situations are inherent in all regulation writing and compliance work. If an item or construction detail is often desirable but not required by law, any contractor would consider its cost prohibitive as far as the final bid is concerned.

In recognition of this limitation on doing more than the bare minimum, regulators have a hard choice when judgment is needed. They often set standards that are more restrictive than necessary in order to cover situations where the item is actually needed. In so doing, they overregulate other jobs. (See Chapter 6 for further comments on this topic in connection with scaffolding.)

OSHA has relatively few compliance officers, and other agencies are similarly limited in their ability to observe every violation. Consequently, safety becomes a business risk. Is it cheaper to risk an accident and the possibility of a fine than to comply with a regulation that seems to be unnecessary? This attitude reduces laws to the status of recommendations or suggestions to be followed if the contractor chooses. Workers quickly sense whether supervisors expect them to take chances and even to ignore the safety rules that were explained to them when they were hired. In a word, overregulation leads to underregulation and disregard of all rules that the individual does not choose to follow.

In some countries, people obey the laws because their families may never see them again if they oppose the authorities, but in America it is impossible to enforce an unpopular law. The British attempts to enforce tax laws led to the American Revolution. Prohibition started the bootlegging industry and gave organized crime its biggest boost. Even the electronics industry has profited from unpopular laws by the sale of radar detectors. But regulators do not seem to have gotten the message; they think that licensing crane operators will stop crane accidents, that licensing engineers will prevent design errors, and that two laws are better than one. Before writing any new law authorities might well give more thought to the enforcement problems it will create. If several individuals fail to live up to their legal responsibilities and an accident occurs, should all be considered guilty or only the person whose action was the direct cause of the accident? Overlapping responsibilities and overly complex laws cause much unnecessary litigation.

In the summer of 1982 a 137 foot tower crane boom was being lowered at the top of a 44 story building when the 70 foot boom of a derrick collapsed and dropped the load onto a granite parapet. Masonry fell to the sidewalk, killing one person and injuring 16 others. The widow of the man who died sued for $50 million, naming the city and three contractors among the defendants. Who caused the accident? Why were so many people sued?

Was the City of New York responsible because it permitted sidewalk sheds to be removed too soon. Investigators recommended that the building code be amended, but the removal of the sheds was legal at the time of the accident.

The engineer who designed changes in the derrick boom may be judged responsible because the part of the boom that buckled had not been reenforced when changes were made to strengthen the boom. On the other hand, the building department's crane and derrick division had approved the changes, welds had been tested, and an inspector had checked the derrick at the site and approved it shortly before the accident. Would it not have been better to have one person responsible, leaving it to his judgment to decide how much checking and testing would be needed?

A case will probably be prosecuted against the rigging foreman for allowing a derrick rated for 22,000 pounds to lift a 24,100 pound load, an overload of 9.5%. The most recent report I have seen indicates that no charges had been brought against him, but that some action might still be taken. I wonder whether it is reasonable to expect every foreman to know weights to an accuracy of less than 10%, and what would have happened to him if he had refused to authorize the boom to be lowered by the derrick?

Possibly the general superintendent on the job, who was the designated safety coordinator should be held responsible because he did not have traffic stopped during the dismantling operation and did not personally check all other details. If a scapegoat must be found, the individual in charge of safety could be the prime candidate. Still, would a jury of his peers be likely to convict him of criminal action? Can one person be expected to be in charge of a project and still give his attention to all details of the safety requirements? For the general superintendent to be the "designated" safety coordinator is to imply that safety matters can be handled in a minimal amount of time; thus he has a title without any attendant duties.

Although in New York City crane operators must be licensed, the licensing procedure could not have contributed to the prevention of this accident. The man at the controls was not really in charge of the operation. Certainly, everybody involved wanted to do his part properly. In the end it was the regulatory system that must bear a large part of the responsibility for failing to prevent the accident.

In summary, it is difficult to provide clear proof that too many rules will actually increase the number of accidents, but excessive regulation does seem to lead to divided responsibility and that inevitably weakens the supervisory process.

2

Planning a Safety Program

2.1 General Strategy

It is only within the last 50 years that business administration has become a recognized profession. Now, safety and loss prevention are beginning to be recognized as important administrative skills, along with accounting, inventory control, marketing, and other business administration specialties. Progress in this area has been faster in some industries than in others. Manufacturers of explosives and most chemicals have been successful in controlling occupational hazards. Workers in many trades formerly considered hazardous are now safer at work than at home. Firms with good safety programs are constantly identifying new problems, and researching solutions to them. Industrial hygiene departments are common in manufacturing plants.

The construction industry has been relatively slow to recognize that safety is capable of contributing significantly to profitability and that it warrants serious attention by top management. I predict that within the next decade or two construction industry safety programs and administration will be significantly different; it is in the interests of all concerned to be sure the changes are for the better. This can be more readily achieved by closer labor-management cooperation than by extension of governmental regulatory activity. The more effective industry becomes in self-regulation, the less pressure there will be for government to interfere.

Although owners of projects are becoming more involved in safety matters, more progress in this direction would be desirable. Hotel chains are beginning to recognize that it is bad for business when people prefer motels or ask for rooms on the lower floors of tall buildings because they are concerned about fire hazards. Similarly, some business firms consider the safety of meeting rooms before making arrangements for assemblies of key personnel. Sports fans have not shown much hesitancy to attend games in new sports arenas, but if there are more roof failures, their concern may increase and seats may be vacant at more events. Owners have, of course, always influenced safety for better or worse. By looking the other way when they knew of safety hazards or code violations, they have encouraged violations. Some catastrophies have begun to make owners see that they do have an interest in safe structures.

Contractors have long known that accidents cut into profits, but they are just beginning to realize that it is possible to make significant reductions in the number and severity of accidents by managing safety in much the same way as they manage other activities.

E.I. du Pont de Nemours and Co. has been a pioneer in safety management since 1903. That company's Construction Division works with their contractors at more than 40 sites in the United States and abroad on projects that have totaled from $600 to $800 million in recent years and have involved 12,000 to 19,000 workers. The safety programs are coordinated by 23 Du Pont and 12 contractor-employed safety engineers who spend 100% of their time as staff advisors and consultants on safety and health activities at construction sites.

Ingo Zeise, Manager of Safety for the Construction Division, outlined the safety program at the fall convention of the American Society of Civil Engineers in 1982. He stressed the importance of management's strong committment to safety, summarizing the policy that "safety has to do with everything and everybody, all the time and under every conceivable circumstance." Proof that the program works is the fact that from 1972 through 1980 the company's incidence rate for lost workdays was 30 to 40 times better than the construction industry average.

The program strives to have every contractor heighten employee safety consciousness from the first day on the job. Each craftsman receives a construction safety handbook that deals with basic safety procedures as part of his safety orientation on the day he is hired. This is followed up by a Safety Task Assignment program. Each time an employee is assigned a new task, his foreman is required to provide detailed safety instruction. Later, the foreman must make sure that the craftsman understands the instructions. It is customary to start every work week with a 15 minute gang meeting on a topic of interest to the individual crew.

The Du Pont construction safety program also includes training of foremen on how to communicate with employees, investigate injuries, and on matters of particular importance to their projects. Specialty training programs are conducted on rigging, erection of scaffolds, crane operation, and fire protection.

Safety inspection tours are made weekly at every site, and deficiencies are with contractor's supervision corrected. If an injury occurs and first aid and medical treatment are given, an investigation into the cause of the injury is conducted, leading to a formal report that is distributed throughout the Construction Division in order to minimize the chance that a similar accident will occur elsewhere.

The safety message is spread at every site by signs, scoreboards, and recognition programs for those who achieve long-term goals. Individual incentive programs are also encouraged.

The cost of administering the program has been compared with losses resulting from injuries in the construction industry as a whole, and Du Pont considers it very cost effective.

New ideas have been accepted by leading contractors and owners, and it is reasonable to expect that they will slowly penetrate the industry. Today's advance thinking becomes tomorrow's normal procedure or is discarded, depending upon the validity of the idea. One new promising development is the increased formalization of safety programs. They are being planned with achievable objectives, means for evaluating progress, and for meeting changing conditions. Recognition of the relation between safety and profits is stimulating competition for better safety records. To achieve a safety record that reduces compensation payments to 20% of the industry average is good, but when a competitor has only a 10% loss ratio and is largely self-insured, he has a definite advantage when bidding.

Some features of safety programs that have been developed by progressive organizations are described in this chapter. Many contractors will probably consider certain ideas too novel or overly structured and therefore unsuitable for their organizations. My challenge to every contractor is to select the good features and adapt the questionable ones to local conditions. More good ideas will be developed in this way than by waiting for the perfect program to come along. We should all live that long! The size of the firm need not be a determining factor in the effectiveness of the program; it is the quality of leadership that counts, and the attitude of the workers. Large contractors have no monopoly on these assets.

A typical general strategy around which a safety program may be structured consists of three main parts:

1. *Motivation.* The art of getting people to want safe working conditions enough to support a program, even when they may not completely understand the reasons for some of its features. Motivation includes confidence in the importance and general validity of the program, and a willingness to participate in making it a success.
2. *Knowledge.* Determining the probable problems and hazards, finding the applicable codes and regulations, choosing effective control methods for minimizing accidents or illnesses, and planning ways to cope with possible emergencies.
3. *Monitoring and adaptation.* Keeping up the momentum of the program, maintaining interest, identifying new problems, changing the program to meet unexpected situations.

In the next sections of this chapter, motivation has been emphasized and considered from a number of points of view. This is the single most important topic in the book; nothing ever happens unless someone wants it to.

In the section about knowledge and awareness I have listed some sources of information and have commented on some of them, both favorably and critically. Later chapters contain more specific reference material to certain hazards. I have also commented on monitoring and adaptation in this chapter and in later chapters where appropriate.

2.2 Motivation

Motivation, as I am using the term, has about the same meaning as leadership. It begins with the study of the way people feel about safety, and how their attitude toward that subject may have been affected by their previous experiences. Furthermore, it is the art of bringing to their attention ideas and facts about their safety at work that will encourage them to avoid accidents. Motivation includes the realization that everyone has problems and conflicting desires; it definitely does not include the concept of manipulation, or any popularized "how to" system that purports to reduce the complex art of management to the level of a do-it-yourself program that anyone can use all the time.

On the other hand, if a safety engineer does not have the knack of getting his ideas across, he may benefit from a safety-oriented seminar in which the emphasis is on seeing one's self as others see one. Malcolm E. Shaw, President of Educational Systems and Designs, Inc., of Westport, CT has conducted safety seminars in which the role playing technique of instruction has been used. A situation is described to two or more participants in the seminar; each is asked to play the part of supervisor, worker, or owner. Others in the group observe how each participant handled himself; they note what he did well, and the ways he could have been more effective. Sometimes the same participants exchange roles and reenact the incident with variations. People who have had field experience and can visualize situations often get very involved in such activity; it can be made very realistic when the problems are well chosen and the participants recognize the possibility that they may have to handle similar situations on the job.

Several good and poor ways of handling a problem may be found, showing that in dealing with people there is not necessarily a right and a wrong method. People are much too complex for such a simplistic notion.

The leader of a seminar may prefer to be called a "facilitator" so that attendees do not look upon him as a lecturer or instructor, but rather as one who is adept at drawing upon the experiences of the group and encouraging participation rather than passive absorption of information.

Simulating situations and exploring alternative ways of coping with problems is also helpful in developing the ability to plan on-the-job meetings, to anticipate problems that may crop up, or to consider ways of handling provocative questions. It is possible to learn these skills on the job, and most supervisors have had to do it that way, but if there is an opportunity to accelerate the process of acquiring experience by any method that allows the trainee an opportunity to correct his mistakes without the considerable difficulties involved in an actual case, the time spent will usually be well repaid.

Everyone is an individual whose moods, attitudes, and dominant motives are constantly changing. It is a legitimate function of management to help workers develop self-control and maturity and to help them gain personal satisfaction from working in a safe manner. If further justification for such motivational activity is needed, I may point out that safety is one field in which the best interests of workers, contractors, owners, and the general public are the same.

Observing how well men are accepting the safety program is important. Everyone who has supervised men knows they respond differently as members of a group than as individuals. A worker may react favorably to an instruction one day and have a negative reaction the next. Skillful supervisors sense when to issue direct orders, exert their authority, and make it clear that compliance is required. They also recognize there is a time to stimulate the self-esteem of workers as individuals, appealing to their pride of workmanship and their status as craftsmen. At still other times, it is better to concentrate on group activity, relying on the cohesiveness of the men who work together to insure the safety of all. The safety engineer needs to sense whether peer pressures are encouraging safe working routines or are undermining his efforts. He should note whether repetition of instructions is contributing to a better understanding of the safety rule or is being resented by the men as unnecessary nagging. Praising an individual in private may be better than where his friends are around; he may not want to be singled out as a special favorite of management.

Special incentives are sometimes considered, such as gifts or bonus pay, for achieving a safety goal. If the safety budget is tight, it is probably better to avoid this method of encouraging cooperation because it may give the impression that management does not put a very high value on safety. Sometimes the solution to this dilemma is to find some item that is novel, not locally available, and still within the budget.

The job schedule is likely to influence the type of motivation that can be considered. Although the safety engineer may want to do a thorough training job, if the project will be finished in a short time, he may have to use an entirely different method of preventing accidents. He may have to concentrate only on the immediate hazards on a day-by-day basis and hope that he does not miss anything.

No discussion of motivation can ignore the attitude of the head of the organization. Without his support, even the best safety engineer will find he is wasting his time. Of course top management is never openly opposed to safety, but there is a world of difference between nominal acceptance and enthusiastic support. Subordinates are quick to observe the sincerity of the boss's committment. If a safety program is nickled and dimed to death, the boss might just as well come right out and say to hell with it.

To clear the air of vague generalities, safety engineers will find a written policy statement from the head of the firm a useful tool. But they may be surprised to find how hard it is to get it written and signed. Federal law requires that there be compliance with OSHA regulations; so a policy statement might logically begin by stating that it is company policy to comply with OSHA regulations. Even this start may be resisted if OSHA citations are being contested or regulations have been ignored for years. It may be easier to get agreement on safety objectives, such a 50% reduction of the accident incidence rate, elimination of disabling injuries, and prevention of fatalities. Aiming for a realistic goal is desirable; it should be challenging but not impossible to reach. The policy statement might say something about the safety engineer's position in the organization and his authority in emergencies. Good policy statements often affirm that regulations are minimum standards, and that cooperation is expected from all workers to exceed the mere letter of the law. One typed page should be ample to state general policy, indicate management's committment to safety, and make it clear that all workers are expected to cooperate with the safety supervisor.

Before the policy statement is released to employees, it should be discussed at a supervisor's meeting where questions may be asked. Try to get the executive who signed the statement to explain it at the meeting; it will help him realize the importance of the paper he has signed and how it is understood by his subordinates. First line supervisors, who deal directly with workers, need to know just where they stand with the safety engineer. They may be apprehensive lest he take away some of their authority over workers or try to countermand their orders. Get these problems out in the open; talk them out to everyone's satisfaction. Unless the people with whom he has to work are motivated to cooperate with him, the safety engineer will never get very far.

Management should recognize that the safety engineer needs a certain amount of status to get results. How he is treated by top management not only affects his own enthusiasm for the assignment, but his influence on others. It costs nothing to include him in plant tours when the owner or other prominent people visit the site.

A good safety engineer carries his weight by anticipating and avoiding troubles, so his contribution may not be as apparent as that of people who solve visible problems. When he solves a problem and prevents an accident, there is nothing to show. But, co-workers will note how the boss listens to safety suggestions, and the extent to which they are implemented, particularly if some expense is involved. Small compliments on a job well done do a lot to keep the safety engineer's head above water in a job that can often be difficult and thankless.

Once the safety engineer and management understand each other, they should review the company's safety record and identify its weak points. How many accidents have been occurring per year? How does the incidence rate compare with the construction industry average? What caused the accidents? Were they investigated in depth, and were steps taken to prevent similar occurrences? Observe the general company-wide attitude toward safety. Are workers and first line supervisors skeptical, indifferent, or enthusiastic about safety? Does anyone else in the company really know anything about safety? One need only read advertisements in the professional employment columns to observe that experience in safety problems is seldom even mentioned by employers or by prospective employees. It seems that few influential people in the construction industry rate safety experience highly. This attitude is shared even by firms that sell safety equipment; in fact, "sell" is the wrong word. Many of them expect the customer to select what he wants with little or no help from the order takers who call themselves salesmen. More enthusiasm and aggressive salesmanship from such firms might make employers more aware of safety hazards, their costs, and the savings a competent safety engineer can make.

A young engineer may find that he has several responsibilities, among which is the safety program. He may find that management's indifference to safety activities indicates general indifference to people. Are employees just people on an organization chart to be moved about as if on a chessboard. If so, how does this affect opportunities for professional development? Discussion of safety policy can reveal the firm's management philosophy: autocratic, authoritarian, paternalistic, team oriented, or focused on objectives to be met as individual key people prefer. Any safety engineer who works for an outfit with a management style diametrically opposed to his own is bound to regret it.

As a result of his initial appraisal of the situation, the safety engineer may well ask himself whether a safety program can be based on the premise that the workers prefer a safe job. Hardly anyone can be found on any job who will publicly express a contrary viewpoint, but what people say and what they truly believe are often very different. A case can be made for the proposition that construction workers, in common with many other typical Americans, are indifferent or even antagonistic toward safety. They particularly dislike safety regulations. Statistics prove that seat belts save lives, but does the typical driver use his? Statistics show cigarette smoking causes cancer and most smokers keep right on smoking. In common with the general population, construction workers are concerned about some hazards but indifferent to others, regardless of statistical findings. They are apt to be more concerned about chemical and electrical hazards than falls, even though falls cause more injuries. Perhaps this attitude is a result of considering falls a form of violence, and we have been conditioned to consider violence a normal condition, an inescapable part of life. For close to a century, almost every family has had someone go to war. Older parents and grandparents still recall World War I, and succeeding conflicts are well remembered. Violence in civilian life is commonplace, and television networks consider it entertainment. The daily news reports condition us to expect accidents to happen. Why, then, should a safety engineer expect that workers will suddenly become acutely aware of hazards as soon as they report for work? Their awareness has been dulled during their off hours, and it is not easy to change attitudes that have persisted for many years.

There are many stable, skilled, mature workers who habitually follow good safety practices. They often do not get the notice they deserve. But proper motivation of the whole work force does include supporting the self-esteem of safe workers, giving them the recognition they have earned. To say, even in passing, "that's the way I like to see a job done" takes only a moment, but cannot help giving the recipient of the praise a better feeling about himself, particularly if others hear the comment.

Occasionally even the most reliable worker can make a mistake and suffer a minor injury. If it was a stupid mistake that caused the accident, he may take a ribbing from his friends and feel that he has to prove how good he is, particularly if they are adept at getting his goat. If he shows off before new employees, who are inherently more accident prone than more experienced workers, they may not realize how risky his display was, try it themselves, and get hurt. An example might be changing a boiler drain cock on a hot steam boiler without draining it or cooling it off. It can be done, but it is very easy to get scalded in the process.

While the mature worker may have occasional lapses during which he fails to use common sense, there are some workers who remain small boys all their lives. They have to walk the shaky planks instead of securing them, climb scaffolding instead of using a ladder, and try to get crane operators to let them ride the hook. Their safety gear is always misplaced or lost. They equate taking chances with manliness and visualize themselves as the successors of the pioneers, cowboys, and stereotyped TV stunt men. Even if they suffer only minor injuries, much managerial time is wasted filling out reports and making investigations. Perhaps this type of juvenile behavior reflects the self-centered life of a small child who is competing with siblings for parental attention. Praising this worker for his performance may give him the recognition he needs, but I seriously doubt it.

Bigots can be a serious problem; they often see the entry of minority group members as a threat to their job security or the solidarity of their group. Even if open conflict does not break out on or off the job, constant needling may cause its victim to do something rash or absentmindedly stumble into a hole. Bigots are almost certain to be antagonistic to any safety program that helps people they dislike to hold jobs without getting hurt. They may be clever enough to avoid doing anything for which they could be disciplined or fired, but they cannot be trusted to report a defective ladder, take a nail out of a board on which someone could step, or do the little things that they would habitually do if working with people they like. If his attacks are purely verbal, aimed at getting his target to quit the job, it may be possible to bring him around, particularly if he can be made to see that such activity stands in the way of an advancement. There is a double motivation problem in situations of this kind. Both the aggressor and his target have to be considered, and both are apt to be tangled bundles of tensions, prejudices, and conflicting motives. For long-term employees such problems have been worked out, but on a construction job that will be completed in less than a year it may be impossible to do more than keep the lid from blowing off.

Alcoholism is a classic result of emotional problems, and no safety engineer can expect to escape problems associated with it, even when the underlying causes are unrelated to the job. After an experience or two, a safety engineer may get into trouble by assuming he is competent to tell whether a worker is intoxicated or not. The National Council on Alcoholism urges all supervisors to treat alcoholism as a disease, and to be extremely careful not to jump to the conclusion that a worker is intoxicated. He may be having a reaction to prescription drugs, suffering from some other illness, or behaving in an erratic manner for any one of many reasons. Supervisors should concentrate on job performance and get the worker to a doctor as quickly as possible. Let an expert determine what is wrong. If the employee is capable of returning to work, no harm has been done. If he is actually drunk or unfit to do his job, no one can argue that management was treating him unfairly. By adhering to this policy of separating responsibility for diagnosis and treatment from responsibility for job performance, much litigation can be avoided.

When a worker is frustrated by inability to resolve a conflict between two or more strong motives, he may find it impossible to pay proper attention to his work. For example, a reliable and competent worker may be offered an advancement that requires him to supervise his old friends. He likes being one of the gang, but he also wants the new job. Meanwhile, he is apt to have an accident or cause one. Even months after he has accepted the new position, the conflict may remain and render him unable to do work for which he is well qualified. An experienced safety engineer may be able to help such a person if his old friends cooperate by letting him know he is still their friend and they are glad to have him as a boss.

Some workers regard safe working methods as personal trade secrets. What they learned the hard way is not to be shared with others. They will use a trick of the trade only when new workers are elsewhere. In a meeting on safety the secretive person may ask questions that he thinks the speaker may not be able to answer, but he will not contribute to the discussion. It is not necessary to be concerned about the personal safety of such people, but it is advantageous to draw them into the program if it can be done without too much effort. If the safety engineer is well qualified to answer questions that are intended to embarass him, so that he gets credit for superior knowledge, the know-it-all will soon get off his back and may even try to garner some glory for himself by participating constructively.

The worker who has a strong craving for attention can be a safety liability. He may bait others by letting them understand that anyone who tries to work safely is a coward. He may boast that he has always done things his own way and never lost a day's work as a result of an accident. Nobody can tell *him* anything about doing his job safely that he doesn't already know. He is likely to talk a lot but avoid doing anything for which he could be discharged. He baits people and, if he manages to create a scene by getting under someone's skin, he has gratified his own ego and made his opponent lose face by allowing himself to be trapped. If this type of person is actually a good worker, he may make a good member of a safety committee and, given some recognition, become an asset.

Pride is a powerful motivator. The image a worker has of himself wearing safety gear can determine his willingness to wear it. Hard hats have come to symbolize virility, association with a rugged group of workers, and someone whose work changes the face of the community. Consequently, they are usually worn with pride. Hard hats come in so many colors and can so easily be decorated with symbols that they are useful in many ways. Having a certain mark on the hat can be prized more than a privilege of tangible value. The requirement that visitors wear hats that are easily recognized may warn workers that visitors cannot be relied on to follow accepted safety rules. Owners, design professionals, and distinguished visitors will probably expect distinctive headgear; they may even bring their own. Making safety belts, welder's goggles, respirators, and indeed any distinctive piece of safety gear a status symbol is a step in the right direction. However, one problem with using pride to motivate workers is that it may conflict with other motives or simply dominate them.

For example, fear and pride may conflict with each other. Construction workers are not timid; they have learned to control their fears, but they are not necessarily unmindful of danger. They just know how to live with it. Bravado may often conceal fear; the safety engineer may recognize it as a front, indicating insecurity and the inability to deal rationally with some hazard. This lack of competence can pose a safety hazard; it reveals the need for training or better engineering control of dangers. But when a safety hazard is reduced or eliminated, the worker may feel that his pride in his ability to cope with that danger is under attack. If so, he will resist the change and may even prefer to take the risk rather than avail himself of the safety measures. Announcing that safety belts, for example, will be issued only to workers whose work is inherently dangerous may help bolster pride and make the belts more acceptable as symbols of the ability to do dangerous work.

Moody workers require more than the normal amount of attention, particularly the person who is "down" more often than usual. He may have temporary problems that distract him from dangers on the job, or he may be suffering the initial stages of mental illness. He may be daydreaming or visualizing himself in another situation. If he does not feel that he is appreciated at home, he may visualize an illness as a means of showing his family how important he is to them. Daydreams can come true, at least to the extent that the illness or accident actually occurs.

I am tempted to investigate more fully the possible relationship between accidents and suicide than the topic probably warrants, but it seems appropriate to comment briefly about it. I can readily understand why doctors are reluctant to report the cause of death as suicide when it may have been accidental. They recognize that the family of the deceased may benefit financially and may have fewer feelings of guilt if the death is reported as an accident. Therefore, unless a suicide note has been found and brought to the attention of the person filling out the death certificate, accidental or natural causes are more apt to be reported.

Statistical data reported by the U.S. Bureau of Vital Statistics and the World Health Organization of the United Nations show that the suicide rate for carpenters, plumbers, and electricians is about twice as high as for the population as a whole. Even this ratio may be low, because there are many opportunities for a would-be suicide to make his death look like an accident.

Some investigators have proposed that there is a definite relationship between changes in job status and suicide rates, but there are many unknown factors. The final decision to end one's life may be delayed, possibly for years after the beginning of this suicidal train of thought. It is possible to get some insight into the state of mind of people who commit suicide by studying people who attempted suicide but were not successful. Nevertheless, there is no way to be sure that these findings apply to those who actually committed suicide. In fact, some psychologists have pointed out that many people who attempt suicide hope that they will fail; if they really wanted to succeed, they would choose other methods.

Safety engineers and foremen should recognize that workers who are going through a major change in their life-style may experience more stress than they think they can handle. If a worker seems unduly depressed and discouraged, he may need professional help. If he does not get the help he needs, an accident may result, and he may cause many deaths besides his own.

Workers are often responsive to incentives that promise future rewards such as advancement or award dinners at the end of an accident-free period of time. They are unlikely to be equally responsive to warnings to avoid dangers that will not visibly affect their health for some time in the future. Asbestos dust is a very real health hazard, but it may not disable a person for as long as 20 years after exposure. More maturity than it is reasonable to expect of many construction workers is required to make them take such hazards seriously. Motivation to conform to regulations may have to be more immediate, such as the possibility of disciplinary action.

Safety engineers may find that their own emotions cause them problems. How can anyone convey enthusiasm for a safety regulation he considers unnecessary? Insincerity has a way of showing itself, regardless of how it is covered over. A little hypocrisy can taint sincere effort. What can be done, for example, when OSHA requires action on the job that seems unsuitable under the circumstances? A regulation should be studied carefully before it is condemned and, if necessary, a call made to the regional office of OSHA for the latest interpretation. If others have voiced similar objections, a new ruling may be in effect. Variances may have been approved, and it may not be difficult to have a better solution to the problem approved. Discussion may even lead to the conclusion that the rule is not too bad after all, or that the difficulty of following it is not sufficient to warrant protesting it. If, after investigation, the rule still seems to be out of line with reality, there may be another way to do the job without involving the regulation at all. For example, if blasting is contemplated but there are too many rules to follow, the same result may perhaps be obtained by using hydraulic jacks or a wrecker's ball. When all else fails, it may be the best policy to admit to all who are involved that you, as safety engineer, consider the regulation too restrictive, but that you are obliged to follow it because it is too much trouble to get a variance. There is no hypocrisy in this attitude, and people who disagree with the rule will understand that sometimes everyone must choose between unattractive alternatives. I do not recommend that any safety engineer sanction disregard of a regulation; if he can disregard it, so can anyone else, and soon there will be no enforcable rules left. Every supervisor and every worker will feel free to choose for himself what safety regulations he should follow.

Safety engineers should follow the same management style when dealing with safety matters as they would when giving orders for doing regular construction work. Safety is becoming a technical specialty; formal courses and home study programs as well as books are written by people who have their own management styles. Inevitably, in talking about safety, some of the instructor's ideas about how to manage workers may influence the foreman's supervisory style. This can cause trouble when a foreman who is used to shouting out orders and chewing out men when they get out of line gets the idea that it is necessary to be a soft-spoken diplomat when he is talking about safety. Workers wonder what is going on, why the difference?

I don't personally consider the drill sergeant style to be the most effective in many situations, but it is often effective and many men like to have it clear who is the boss. They don't want to be asked to help make decisions. If a foreman is in earnest about safety, it doesn't matter how he gets his men to avoid accidents, provided he gets results.

The amount of detailed instruction given to workers is also observably different from one foreman to another. One may spell out all of the details and another may leave it largely to his crew to work them out. The same patterns apply to safety rules. One foreman may simply say that he wants all ladders to be secure before they are climbed, and leaves it to his workers to decide whether to block them, tie them, or use a job-built ladder and nail it in place. Another foreman may spell out every detail. Both systems can work, but it is confusing if the "general instruction" foreman suddenly becomes a "detailer" when it comes to safety.

I have already mentioned the importance of compatability between the management styles of safety supervisors and other managers. The same principle applies to communication. If the organization wants all communication to be in writing, even a safety engineer who prefers a less formal communication system can generally adapt himself to that organizational preference. However, the important consideration is communication. It is not inherently better that everything be in writing; sometimes the less there is to read the easier it is to find the important points. A safety engineer who walks the job often and simply calls attention to any hazard that should be changed may be more effective than one who writes a lot of survey reports. He may not even have to say anything; kicking a loose plank or bending over a protruding nail may be sufficient to indicate that he wants things set right. If he gets his message across, he does not need to write reports that detail poor housekeeping or list violations of safety rules. When the time to write a report comes, he can be lavish in his praise of the way foremen are doing their jobs. If body language has been used to indicate all the bad comments, verbal or written reports can still record the complimentary items.

Emphasis on the positive is contageous. When a foreman has obvious pride in the safety record of his crew, they are more likely to live up to his expectations than if he is constantly nagging them to be more careful. If the same precautions are repeated too often, they can go in one ear and out the other without encountering the brain at all.

Investigation of near misses is important in making workers aware of hazards. An accident that almost happened is like a rehearsal for one that is about to happen. If the rehearsal is ignored, the golden opportunity to prevent the actual accident is lost. Near misses are good topics for safety meetings. They provide concrete examples for discussion. Causes and possible consequences can be discussed without the limitations that may follow an actual accident, such as the absence of the injured worker. Furthermore, near misses are likely to be far more numerous than actual injuries. If near misses are substantially reduced, the number of injuries will probably be reduced in about the same proportion. The concern of management with accident prevention will be emphasized to workers if they realize that it is not necessary for anyone to get hurt before a hazard is investigated. *Pre*vention implies action *pre*vious to an accident; what better occasion can there be for such preventive action than an accident that almost happened.

When the job has been completed, if the end result is an excellent safety record, the motivation problem may have been solved in any number of ways. It is the end result that counts, not the means chosen to reach it. The difference between a good solution and an excellent one is the way the safety program has supported the other supervisors' activities. When workers feel good about themselves when they are following safety rules, they probably feel equally proud of the way they are producing. It can all fit together if motivation is skillfully handled.

2.3 *Knowledge and Awareness*

Knowledge and awareness are closely related. Everyone has found himself suddenly aware of the present importance of something he once knew but had forgotten. Sometimes a minor injury reminds us that we made the same mistake before, and we wonder why we were foolish enough to make it again.

Simply thinking about safety will often remind us of useful ideas we formerly found applicable, or accidents we avoided because we were lucky or someone warned us in time. No one has to start from scratch in his search for useful knowledge about safety; we all know more than we are likely to realize. However, there are times when new problems have to be faced and our experience is inadequate. Then, the following suggestions may be useful. Appendix K, a checklist, will undoubtedly remind every safety engineer of incidents from his own experience that he has not recalled for years.

So much has been published about safety that anyone who attempts to dig out information about a specific safety hazard should be prepared to find hundreds of pages of material about it, but much of it will be repetitious or lacking in substance. Some publications will be obscure; standards and regulations may include technical terms or redefine words that are in common usage to give them a specialized meaning. For example, gases, fumes, vapors, and mists are more or less alike as far as most people are concerned, but an industrial hygienist must consider them to be technically distinct forms of matter. One must learn which terms are significant in each specialty, just as every contractor knows the importance of the difference between "shall" and "should." Getting started is, therefore, often the most difficult part of an investigation. I have tried to address this problem throughout this book, assuming that if a safety engineer is given some help in locating some basic information he will have no difficulty following up the details.

One good starting point is often the U.S. Government Code of Federal Regulations (CFR). It is divided into many parts and subparts and, therefore, the first step is to find a few references to the general field in which one is interested. After that, it is relatively easy to find out more about related matters; information will probably be in the same part of the CFR. I have therefore given references to parts or specific paragraphs in the regulations throughout this book. Even if the reader does not care to look up the references himself, he will find that specialists with whom he may discuss problems will probably be familiar with the referenced paragraphs. For example, in Chapter 3, Section 3.3.3, I mention 29CFR 1926.52 (d) (1) and reproduce a brief table dealing with noise and hearing protection. Every expert in the field of hearing protection will be familiar with that reference and will probably be able to pull a copy from his files or pocket notebook. Furthermore, the CFR paragraphs often mention other sources of information. In Appendix G, on respirator selection, I mention 29CFR 1926.103, and anyone who reads that portion of the regulations will observe that it is followed by additional information on the same general subject, including mention of the specification for Grade D breathing air, as described in Compressed Gas Association Commodity Specification G-7.1-1966. With that hint, a safety engineer will deduce that a call to the Compressed Gas Association could be productive, and he is well on his way to finding out whatever he may need to know about respirators. There are many ways to get started on an investigation; I have tried to provide a few that will lead to others.

I have said little about industry standards, labor laws, and local code requirements. Contractors will probably get more up-to-date information about such matters from manufacturers, labor leaders, and their own association officers than from any book. Interpretation is very important when this type of material must be used and any reading should be supplemented by discussions with people who are familiar with the details.

Any comprehensive list of organizations in the construction safety field and related activities would be cumbersome to use; it would necessarily include hundreds of entries, and for each entry there could be additional entries that were related to the basic groups in some way. As the list became longer it would become less useful. Therefore, I have prepared notes on only 13 sources of information, as described in the following sections:

- 2.3.1 Occupational Safety and Health Administration (OSHA)
- 2.3.2 National Institute for Occupational Safety and Health (NIOSH)
- 2.3.3 Superintendent of Documents, U.S. Government Printing Office
- 2.3.4 National Technical Information Service (NTIS)
- 2.3.5 National Safety Council
- 2.3.7 Building Trades Unions
- 2.3.8 Insurance Companies
- 2.3.9 National Fire Protection Association
- 2.3.10 Business Periodicals Index

2.3.11 Local Regulations
2.3.12 Manufacturers
2.3.13 British Health and Safety Executive

2.3.1 Occupational Safety and Health Administration (OSHA)

Public Law 91-596 (OSHA publication 2001) is the Occupational Safety and Health Act of 1970. It is a small pamphlet of only 31 pages and is surprisingly easy to read. Copies are available from the Superintendent of Documents, U.S. Government Printing Office, Washington, D.C. 20402, order number 022-003-99902-7. This law has had so much influence on industrial safety that I would put it at the top of the list of recommended reading for anyone interested in safety. Prior to its passage, this legislation was strongly advocated by labor leaders; spokesmen for industry were generally less enthusiastic. Contractors should note section 11 on judicial review, in which paragraph c protects a worker against discrimination if he exercises his rights. When anyone refers to a worker's rights under "11c," this is the origin of the expression.

In the section on penalties, number 17, violations of certain rules may lead to criminal action. Violation of regulations leading to the death of a worker, giving advance warning of an inspection, and giving false information are in this category. Of all the penalties listed, only the advance warning and false information provisions are applicable to workers. All the rest apply only to employers. Every employer should be familiar with this section of the law.

The law created the Occupational Safety and Health Administration (OSHA). This agency has promulgated so many regulations that just about everybody is confused when first exposed to them. I found it helpful to read "All About OSHA," a booklet that is available free from OSHA regional offices (listed in Appendix B). This publication summarizes employer responsibilities. I have reproduced the list in Table 2.1.

Note that the first item in the table requires that employers comply with specific regulations; it is not sufficient that he simply rely upon his own good judgment to provide a safe workplace. Therefore, every contractor should have a manual of regulations applicable to construction. I recommend the OSHA subscription service, Vol. III. Subscribers receive a basic manual of about 600 pages, and thereafter they receive timely supplementary information about changes and interpretations of regulations. Everything is punched for insertion in standard three-ring notebooks, which is a great convenience over other presentations of the same material. Write the regional office of OSHA for current price and ordering information on *Construction Industry Standards and Interpretations*, Vol. III, OSHA 2079. Libraries may have OSHA regulations in book form, published as

TABLE 2.1. Employer Responsibilities

In "All About OSHA," it is stated that as an employer you must:

[a] 1. Meet your general duty responsibility to provide a workplace free from recognized hazards that are causing or are likely to cause death or serious physical harm to employees, and comply with standards, rules and regulations issued under the Act.
2. Be familiar with mandatory OSHA standards and make copies available to employees for review upon request.
3. Inform all employees about OSHA.
4. Examine workplace conditions to make sure they conform to applicable standards.
5. Minimize or reduce hazards.
6. Make sure employees have and use safe tools and equipment (including personal protective equipment), and that such equipment is properly maintained.
7. Use color codes, posters, labels or signs to warn employees of potential hazards.
8. Establish or update operating procedures and communicate them so that employees follow safety and health requirements.
9. Provide medical examinations when required by OSHA standards.
10. Report to the nearest OSHA office any fatal accident or one which results in the hospitalization of five or more employees.
11. Keep OSHA-required records of work-related injuries and illnesses, and post a copy of the totals from the last page of OSHA No. 200 during the entire month of February each year. (This applies to employers with 11 or more employees.)
12. Post, at a prominent location within the workplace, the OSHA poster (OSHA 2203) informing employees of their rights and responsibilities. (In states operating OSHA-approved job safety and health programs, the state's equivalent poster and/or OSHA 2203 may be required.)
13. Provide employees, former employees and their representatives access to the Log and Summary of Occupational Injuries and Illnesses (OSHA No. 200) at a reasonable time and in a reasonable manner.
14. Cooperate with the OSHA compliance officer by furnishing names of authorized employee representatives who may be asked to accompany the compliance officer during an inspection. (If none, the compliance officer will consult with a reasonable number of employees concerning safety and health in the workplace.)
15. Not discriminate against employees who properly exercise their rights under the Act.
16. Post OSHA citations at or near the worksite involved. Each citation, or copy thereof, must remain posted until the violation has been abated, or for three working days, whichever is longer.
17. Abate cited violations within the prescribed period.

[a]Item numbers do not appear in the OSHA publication.

Title 29, Code of Federal Regulations, Parts 1910 and 1926, but these bound volumes may not be up to date.

In Table 2.1 the first two items mention "standards" that must be met. Some are OSHA standards, but many were issued by other governmental agencies, technical societies, or trade associations. OSHA regional offices should have a reference copy of every standard mentioned in the regulations, and they should also know where an employer may obtain a copy for his own use.

To illustrate how standards have been integrated into regulations, consider subpart U of 29CFR 1926, which deals with explosives. In paragraph 1926.914(a), a table of distances prepared by the Institute of Makers of Explosives is mentioned. In paragraph 1926.904(a), the requirements for the storage of explosives in accordance with Internal Revenue Service regulations in 26CFR181 are given. Transportation of explosives is governed by Department of Transportation regulations in 49CFR. Therefore, if every detail of an OSHA regulation must be fully understood, it may be necessary to examine several references. Manufacturers are usually well informed about the regulations that affect their products, they should be able to provide photocopies of pertinent paragraphs. If a regulation is hard to understand, the manufacturer probably knows as much about its actual meaning as anyone else.

Item 3 in Table 2.1 requires that a poster, OSHA 2203 be displayed at each job site. Employees will find it answers most of their questions, but if they want to read any of the actual rules they have a right to see them.

The rest of the items in the table are either self-explanatory or are the subject of booklets available through OSHA.

While Table 2.1 does not specifically state that employers are responsible for training employees, many regulations limit job assignments to persons who are "certified," "competent," or "qualified." Sometimes, as in the case of scaffolding design, the qualifications required are spelled out: a qualified engineer competent in the field may design wood-pole and tube and coupler scaffolds within certain limits, but only a licensed professional engineer may write specifications for tubular welded frame scaffolds over 125 feet high. Contractors should therefore note regulations for specific training requirements that must be met. In case of an accident, investigators will probably check this point very carefully.

Contractors need not assume all responsibility for extensive training of workers; applicants are presumed to be qualified to do their jobs at the time they are hired, with certain exceptions. However, some specialized training may be necessary and records of all such activity should contain the names of those instructed as well as the material covered. Be sure to include notes on safety matters. If an accident occurs, an employer will be expected to demonstrate that he complied with training requirements that include accident prevention.

OSHA publication number 2254 is a comprehensive summary of training requirements, and anyone interested in that subject should get in touch with OSHA's Office of Training and Education, U.S. Department of Labor, Washington, D.C. 20210 or one of the regional offices listed in Appendix B.

2.3.1.1 ENFORCEMENT POLICIES AND COMMENTS ABOUT OSHA. OSHA got off to a bad start in the early 1970s. Too many regulations were issued that dealt with insignificant details. Many regulations have now been simplified, and training of compliance officers has been improved. The OSHA Review Commission and the courts have cleared up many ambiguities. Unfortunately there has been little or no dramatic improvement of the safety record of the construction industry and, until that objective is achieved, OSHA cannot be considered a success.

I think that OSHA's use of the surprise visit technique to detect violations of regulations (unless there is reason to suspect that a contractor is deliberately attempting to conceal unsafe practices) is an impediment to cooperative effort. OSHA is trying to encourage contractors to follow regulations, but at the same time there is the threat of a visit that will disrupt operations, take supervisors away from previously planned activities, and possibly result in citations that may or may not be sustained if appealed. This enforcement technique implies that all or most employers are trying to benefit at the expense of workers, and this is an unjustified assumption.

Contractors are accustomed to complying with plans and specifications that indicate the end result desired, but they expect freedom to use appropriate means of their own choosing to accomplish the owner's objectives. Some OSHA regulations infringe excessively on this freedom of action. For example, standard guardrails are required on scaffolds over four feet high if their minimum width is less than 45 inches [See 29CFR 1926.451(a)(4)]. No doubt there are jobs where such railings should be provided, but certainly there are many jobs where common sense indicates that little or no improvement of conditions would result from the use of the railing. The regulation is so generally violated that it weakens respect for other rules. In a similar category, I would place 29CFR 1926.450(a)(10), which reads "Portable ladders in use shall be tied, blocked, or otherwise secured to prevent their being displaced." I agree that ladders should often be more securely placed and that it is a good idea to tie them to a secure structural member more often than is customary practice, but the arbitrary "shall be" rule is certainly overkill. It puts authority in the hands of a compliance officer in excess of the degree most people would approve and, when bad blood develops between an inspector and a contractor, the inspector has the power to throw the book at the contractor, causing him unjustified expense for appeals. In such cases, there is need for OSHA to display more respect for the integrity and good intentions of the contractor.

At the same time, contractors should recognize the difficulty of writing regulations that are both reasonable and enforceable. A recent case involved a citation for failure to provide workers with safety belts and lifelines while they were installing prestressed concrete panels in a multistory building. The contractor claimed that no specific rule applied and that he was conforming to the customs of the industry. He appealed to the OSHA Review Commission, which ruled that the citation was appropriate because a reasonable person familiar with the situation would recognize the hazard. The case was brought to the U.S. Court of Appeals for the Fifth Circuit, which decided that OSHA could not use the "reasonable person" principle to sustain a violation of a general safety rule, but was obliged to cite a specific rule or industry standard. This decision protects a contractor from arbitrary citations based on a compliance officer's unsupported opinion that a safety hazard exists, but at the same time it makes it necessary for OSHA to write very rigid regulations in order to make them enforceable. As long as OSHA policy emphasizes enforcement instead of persuasion, a considerable amount of animosity is to be expected between the agency and the contractors whose safety records it hopes to improve. I see no easy solution to this problem, but it is certainly one to be reckoned with.

Another stumbling block in the path toward better OSHA–Contractor relations has to do with the responsibility for compliance of workers with all regulations. OSHA is aware of resistance to using hard hats, respirators, and other safety gear if it is uncomfortable to wear, yet full responsibility for not only supplying the gear but also for requiring that it be used is placed on the employer. It is not easy to meet this obligation if the employees have good reasons for objecting to wearing the equipment. Before wearing it, they want to know that the need for it has been clearly established under local job conditions. For example, the OSHA definition of an enclosed or confined space includes open top pits more than four feet deep (see Section 7.7). If an open pit may contain hazardous gas, the safe procedure is to require workers to wear respirators, in anticipation of possible trouble. Workers may feel that it would be better to wait until there is a proven need for protective gear before requiring that they wear it.

Workers recognize that the government has the right to require the installation of sprinkler systems, for example. They feel differently about rules that tell them how to do their jobs and are likely to reject the idea that any government employee knows how they should do their work or what they should wear. Making employers responsible for enforcing rules that workers don't consider necessary and that foremen also think are useless can create real problems for safety engineers.

Here is what section 5 of the Act itself has to say about "Duties."

(a) Each employer—

(1) shall furnish to each of his employees employment and a place of employment which are free from recognized hazards that are causing or are likely to cause death or serious physical harm to his employees;

> (2) shall comply with occupational safety and health standards promulgated under this Act.
>
> (b) Each employee shall comply with occupational safety and health standards and all rules, regulations, and orders issued pursuant to this Act which are applicable to his own actions and conduct.

Note that in both paragraphs 5(a) and 5(b) the word "shall" is used, implying an equal degree of responsibility for compliance with the law on the part of employees as well as employers.

Section 17 of the law deals with penalties, but the only acts for which an employee may be penalized are giving false information or warning an employer that an inspection visit will be made.

I wrote the Assistant Secretary of Labor in charge of OSHA and received a comprehensive response from the Acting Director of Federal Compliance and State Programs, explaining that the question of employee responsibility for compliance had been examined by the courts and that total responsibility rests on the employer. Policy was summarized in OSHA instruction STP 2-1.101 issued July 22, 1981, of which paragraphs E and F are particularly interesting:

> Par. E: Background. 1. Section 5(b) of the Occupational Safety and Health Act of 1970 states that employees shall comply with the standards, rules, regulations, and orders that pertain to their own actions or conduct. However, there is no provision in the Act for direct enforcement of this obligation. On the contrary, the Act specifically places upon the employer the responsibility to assure safe and healthful working conditions.
>
> Par. F. Guidelines 2. For the purpose of this instruction, it is prohibited to apply a sanction against an employee who removes or otherwise disregards an agency order (the so-called "red-tag" provisions) to close down dangerous equipment or worksites which constitute an imminent danger to employees. The prohibition of this type of employee sanction is based upon the premise that it is the employer's responsibility to assure employee compliance with "red-tag" provisions just as it is the employer's responsibility to assure employee compliance with standards.

In view of the foregoing indication that OSHA will not provide the employer any legal support in enforcing safety rules, contractors may find it advisable to keep employees aware of the part of OSHA publication 2253 "Workers Rights Under OSHA" that reads, on page 6, "You are expected to comply with all internal or administrative safety and health regulations issued by your employer, including the wearing or using of protective gear and equipment. Failure to do so may result in your employer's reprimanding, suspending, or dismissing you."

This admonition, tucked away in a pamphlet about employee rights, is discouragingly weak support when employees rebel against rules the employer is obligated to enforce.

OSHA also has a serious enforcement problem; too many job sites and too few people to visit them. Failure of surprise visits to create enough fear to assure compliance must have been a disappointment to OSHA. The first tangible evidence that changes were planned was the OSHA sponsored consulting service intended primarily for small contractors. On request, a consultant who is not one of OSHA's compliance officers will assist an employer in locating hazards and planning ways to eliminate them. Consultants are paid by OSHA, but they do not have authority to issue citations. However, if a serious hazard exists and it is not eliminated within a time considered sufficient by the consultant a visit by a regular compliance officer may be scheduled. The employer may limit the consultant's visit to part of his project, but he has no voice in the selection of the consultant.

These features of the consulting service favor the contractor who can afford to retain a safety engineer of his own choice and pay the fee himself. He can then control the distribution of the consultant's report and, if he does not find the conclusions acceptable, he can debate the matter or call in another consultant for a second opinion, all without the threat of a surprise visit being initiated by the consultant's work.

OSHA has also been experimenting with a few cooperative programs that exempt certain carefully selected contractors from surprise visits if they have a good safety program and a good safety record. Labor-management committees are an important feature of such programs. It will be interesting to see whether these experiments make further significant reductions in safety records that are already good, but I question whether they will have any major effect on the overall safety of the construction industry. The outfits that need the most encouragement to improve their safety records do not qualify for participation in these cooperative programs. Still, they seem to be a move in the right direction and may help make contractors realize that they need to develop their own safety programs, consistent with their own management styles. This movement would probably be given a boost if OSHA would put more effort into the dissemination of information about practical methods of reducing safety hazards, ideas that have been proven in the field to be cost effective and acceptable to both employees and employers. In America, enforcement of a regulation that does not have broad popular support of the people regulated has little chance of being effective. I cannot recall any successful government program that tried to force people to do something for their own good when they themselves did not want the government to interfere. It may be that if contractors and workers are given realistic information about why accidents occur and how they can be avoided, effective programs will be a natural rather than a forced development.

OSHA will provide a speaker for any group of about 15 or more people that would like to know more about safety in the construction industry. In general, at least one hour should be allocated for the presentation, and it could be as long as a full day seminar. The availability of speakers and details of current topics that might be covered can be obtained from the regional OSHA offices listed in Appendix B. Meetings are usually held away from any construction site, and therefore the speakers do not make any observations of on-site conditions; their objective is to explain policies and to answer questions. This is a service that should be encouraged by contractors, and brought to the attention of the program chairmen of organizations whose members are interested in the industry.

Does this mean that OSHA should have no enforcement powers? I hardly think such an extreme position would be realistic, but concentrating on serious hazards should be sufficient. Even elimination of the surprise visit enforcement technique would not leave workers without protection because they already have the right to request an investigation of a hazard and are protected against discrimination for exercising their rights. With greater participation of workers in safety programs, the employer's enforcement difficulties should be reduced. I do not pretend to have all the answers to construction safety problems, but when present methods fail to produce results, the search for better ones must go on. OSHA has the potential to make a significant contribution; whether it lives up to that potential remains to be seen.

2.3.2 National Institute for Occupational Safety and Health (NIOSH)

This institute was established by section 22 of the Occupational Safety and Health Act of 1970 to develop and establish recommended occupational safety and health standards, to investigate toxic substances, and to perform or sponsor research intended to improve occupational safety and health.

NIOSH reports are usually technical and are apt to be of most interest to people who are trying to develop standards or to find out more about the severity of unknown hazards. NIOSH and OSHA have similar objectives, but NIOSH is part of the Department of Health and Human Services (formerly the Department of Health, Education and Welfare) whereas OSHA is in the Department of Labor.

The extent of NIOSH's activities may be judged by examining its voluminous *Publications Catalog, Publication 80-126*. Headquarters of the publications dissemination activity of NIOSH is at 4676 Columbia Parkway, Cincinnati, Ohio 45226.

When a NIOSH publication first becomes available, an individual may generally obtain a copy free from NIOSH, but after the number of copies allocated for such distribution has been exhausted they may be available only from the Government Printing Office (GPO) or from the National Technical Information Service (NTIS). See Sections 2.3.3 and 2.3.4 below for further information about these agencies. Both agencies charge for NIOSH publications; the Government Printing Office price may be very substantially less than the charge made by NTIS for the same publication.

I suggest that before ordering NIOSH reports that an inquiry be made at the nearest Federal Depository Library. These libraries are apt to have reports that are of interest to contractors and, if it is not readily available, it is probably of interest only to researchers.

2.3.3 *Superintendent of Documents, U.S. Government Printing Office*

The address of this agency is Washington, D.C. 20402 and at the time of this writing about 20,000 publications are available. Among them are a monthly index of U.S. Government publications and many Subject Bibliographies. One Subject Bibliography, SB 216, on the Construction Industry, contains 29 pages of titles, order blanks, and prices.

The Superintendent of Documents has a close working relationship with 1350 Federal Depository Libraries located in all the states. Of these, 49 are Regional Depositories where all publications are available; the rest have only selected material, and the topics which may be covered will differ from library to library. To get a list of all the Federal Depository Libraries, write the Superintendent of Documents; it is free.

Contractors should make more use of this valuable resource. If it is possible to travel to one of these libraries without too much difficulty, the research librarian will probably save the visitor many hours that would otherwise be wasted writing to agencies without knowing exactly what title to order. They often have a copy of the construction industry bibliography and may even have copies of some of the items listed in it.

2.3.4 *National Technical Information Service (NTIS)*

Over 1.2 million technical reports are available from NTIS, which operates under the administrative direction of the U.S. Department of Commerce. The service is self-supporting and charges for all technical material. The collection of documents is constantly increasing, and any title in the collection may be obtained for an indefinite time because after the original documents have been distributed, photocopies are made from file copies that are retained permanently. Unlike material compiled by other agencies, such as NIOSH, NTIS collects and distributes reports from all governmental agencies, private industry, and many foreign countries.

Contractors will seldom find it necessary to use this source of information, but when an unusual safety problem arises, there may be no other convenient way to obtain material that has been published about it. Free information about NTIS, including an outline of its activities, may be obtained by writing the U.S. Department of Commerce, National Technical Information Service, 5285 Port Royal Road, Springfield, Virginia 22161.

2.3.5 *Department of Transportation (DOT)*

The regulations governing transportation of hazardous materials such as flammable liquids and explosives are in Title 49 of the Code of Federal

Regulations (49CFR), and I have found that information on regulatory matters was readily available from DOT by telephone.

The Association of American Railroads, Bureau of Explosives, publishes a loose-leaf manual, reprinting DOT regulations as soon as they are released or revised, and they cover all modes of transportation (air, highway, and rail). Most carriers subscribe to this service and at the time I am writing, the latest edition is called "Tariff No. BOE-6000-B." For more information, write the Association of American Railroads, Bureau of Explosives, 1920 L St., N.W. Washington, D.C. 20036.

International agreements have recently been reached that simplify packaging requirements for firms engaged in foreign trade. When containers are standardized, it is reasonable to expect that the metric system of weights will be used. Safety engineers will need to reeducate people who handle materials so that cranes and scaffolds are not inadvertently overloaded. There is always the danger that a worker may forget that 1 kilogram weighs 2.2 pounds, but it is even more dangerous if he gets the conversion backwards and thinks that 1 pound weighs 2.2 kilograms. In that case, he would think that a 1000 kg lift weighed less than 500 pounds instead of its actual weight of 2200 pounds.

2.3.6 National Safety Council

The National Safety Council is a nonprofit, nongovernmental privately supported safety organization, founded in 1913 and chartered by the Congress of the United States in 1953. Employers, insurance companies, schools, other professional societies, labor unions, and a variety of other types of organizations may become members of the Council. Contractors will find the wide variety of services available to be interesting, and I suggest they write to 444 N. Michigan Ave, Chicago, Illinois 60611 for a copy of their *General Materials Catalog*. Industrial Safety Data Sheets are available on more than 300 topics, booklets, and leaflets suitable for distribution to workers, and training aids such as slide and film programs are listed. Many companies distribute *Family Safety Magazine* to their employees, mailing it directly to their homes. It is a quarterly publication featuring timely safety topics, and it seems to help support the company program when workers' families are also interested in safety. The National Safety Council's activities span all types of safety problems; it is therefore difficult to summarize their scope other than to say that anyone having a safety problem is likely to find someone on the Council's staff who has had experience with the same difficulty.

2.3.7 Building Trades Unions

Labor unions have become aware of their obligation to be of service to their members in many ways other than as participants in union-management bargaining sessions. Training in safe working methods is one of these benefits provided by some unions. Because workers continue their union affiliation as they move from job to job, the union is in an ideal position to offer safety training with a greater degree of continuity than is available to nonunion workers. Training courses and safety aids provided by certain unions were summarized in a *Guide to Worker Education Materials in Occupational Safety and Health: Volume II,* prepared under a contract grant from the U.S. Department of Labor, OSHA.

Some training aids and the AFL-CIO unions sponsoring them are as follows.

1. Building and Construction Trades Department, Safety and Health Education and Training Department, 815 16th St. Washington, D.C. 20006. Eleven courses applicable to construction safety are offered, and a book was sponsored *Guidebook to Occupational Safety and Health.* The book was published by Commerce Clearing House, 4025 W. Peterson Ave., Chicago, Ill. 60646.
2. The International Association of Heat and Frost Insulators and Asbestos Workers, AFL-CIO has prepared a leaflet on safe practices for the removal of asbestos.
3. International Brotherhood of Boilermakers, Iron Ship Builders, Blacksmiths, Forgers, and Helpers locals have been pioneers in shipyard safety, and much of their work on materials handling, heat stress, welding, and safety committees is of interest to other trades.
4. International Association of Fire Fighters, 1750 New York Avenue, N.W., Washington, D.C. 20006 publishes a monthly magazine, *The International Fire Fighter;* it contains articles on safety and health.
5. International Brotherhood of Painters and Allied Trades of the United States and Canada 1750 New York Avenue, N.W., Washington, D.C. 20006 has a comprehensive videotape library; tapes on many safety topics are available for sale or rent. They also have two textbooks dealing with hazards and OSHA regulations of interest to their members.
6. Operative Plasterers' and Cement Masons' International Association of the United States and Canada, Safety and Health Program, 1125 17th St., N.W., Washington, D.C. 20036 has pamphlets, slide shows, and transparencies about safety, with particular emphasis on ladders and scaffolding.
7. United Association of Journeymen and Apprentices of the Plumbing and Pipefitting Industry of the United States and Canada, Department of Safety and Health, 901 Massachusetts Ave., N.W., Washington, D.C. 20001 has published some pamphlets about the hazards of working in confined spaces and several other topics of interest to their members.
8. United Rubber, Cork, Linoleum and Plastic Workers of America, Office of Industrial Hygiene, 87 South High St., Akron, Ohio 44308 has published a series of pamphlets about chemicals to which their workers

might be exposed. Some of this information could be of interest to installers, although the emphasis appears to be primarily on the manufacturing of linoleum and plastic products.

This list is just a sampling of the available material. I suggest that contractors discuss safety with the officials of the local unions and ask them what their national headquarters may have prepared. Union publications are likely to be realistic and to the point, based on practical experience.

2.3.8 Insurance Companies

Safety is profitable to insurance companies; it is to their advantage to help their customers avoid accidents. I do not mean to infer that their interest in safety is totally devoid of human interest, but they do have more facts available about the costs of accidents than any other group of companies. They are in a good position to help contractors get the most for money spent on safety programs. It makes no sense to squander the safety budget on activities that have no significance when a good return on the investment might just as easily be obtained.

Wausau Insurance Companies, Wausau, Wisconsin 54401 has done a great deal of work to help employers minimize losses. For many worker's compensation classifications, they have tabulated the major causes of injuries and the costs associated with those causes. For example, for classification number 5190, Electrical Wiring, their study covered 1519 injuries costing nearly $6 million. Electric shock was *not* a major cause of injury. The training budget for preventing these injuries would be more appropriately spent helping electricians to avoid falls and back injuries. Thus, by focusing attention on the major causes of losses, Wausau has made an important contribution to the advancement of construction safety.

The American Mutual Insurance Alliance, 20 N. Wacker Drive, Chicago, Illinois 60606 has prepared some useful leaflets about specific hazards. They come right to the point in a few brief paragraphs. Consequently these publications are appropriate adjuncts to tool box meetings.

The American Insurance Association, 85 John Street, New York, NY 10038 publishes both technical and popular types of booklets. The association has a manual of about 400 pages for those who are preparing for the Certified Safety Professional examinations.

These are just examples of materials offered by many insurance carriers. Contractors should consider these fringe benefits when choosing between firms that would like to write policies for them.

2.3.9 National Fire Protection Association

The National Electric Code is sponsored by the National Fire Protection Association under the auspices of the American National Standards Institute, but this is just one of many activities of NFPA. Since 1896 it has been active in developing methods of reducing the hazards of destructive fires. Membership is open to individuals as well as to organizations. The NFPA has over 800 fire safety publications, including codes, standards, books on fire prevention, and visual aids. Members receive a newsletter and may subscribe to several periodicals.

For details, write to the association at 470 Atlantic Avenue, Boston, MA 02210.

2.3.10 Business Periodicals Index

This index is published by H.W. Wilson Company, 950 University Avenue, Bronx, New York 10452. It lists the technical and business-related articles in over 200 magazines, classifying them under headings such as "Occupational Health and Safety" with many cross-references to related topics. Monthly paperback indexes are consolidated annually; the typical annual index contains about 1500 pages and covers many thousands of articles. This index is particularly useful to anyone who wants to get a sampling of what has been written about a new development in the safety field, or articles about a hazard such as asbestosis. It is far easier to check two or three annual volumes of the index than to thumb through the table of contents of each issue of each of a dozen or more magazines in which articles of interest might be expected to have appeared. Unlike other indexes to periodical literature, the *Business Periodicals Index* does not cover popular magazines in which articles are likely to be overly simplified for the general public.

2.3.11 Local Regulations

States, counties, and municipalities very often have regulations that supplement federal laws. For example, the transportation of hazardous materials is governed by the federal Department of Transportation, but their movement in and out of cities is often regulated by local laws. Shippers must abide by federal as well as local regulations. Any contractor who anticipates doing work overseas should pay particular attention to problems that may arise if he attempts to ship any hazardous materials by air. He has to comply not only with the U.S. Department of Transportation regulations, but with those of the International Air Transport Association and of the individual airlines. Finally, after all these rules have been satisfied, the pilot of the plane may arbitrarily refuse to accept the shipment. This is an extreme example of the complications that may be encountered, but local authorities are often very jealous of their prerogatives. It is not possible to even list the local safety rules that a contractor may encounter. I can only suggest that their existence be investigated before any job is started.

2.3.12 Manufacturers

Manufacturers are an excellent source of information about safety precautions and regulations affecting the transportation and installation of their products. Remember that they get all the complaints; they know what may go wrong. Consequently, reputable firms are very anxious to avoid problems. They are apt to have the most up-to-date information on codes and safety regulations. Some companies will train mechanics; others will supply training manuals or leaflets that are useful at tool box meetings. If a regulation is difficult to interpret, they probably know the answers and, if they do not, a person on their technical staff may be a member of a professional society and know someone well qualified to supply the necessary information.

Contractors should never hesitate to ask their suppliers for help in anticipating or solving safety problems. If a manufacturer is on his toes, he will welcome a chance to help any customer; it is the most effective form of salesmanship. On the other hand, if a request is ignored or poorly handled, the experience is also valuable. The firms that miss opportunities to help their customers are the ones that lose interest as soon as an order is signed. I have found them to be the ones who are slow with their deliveries and impossible to deal with when a complaint must be handled.

2.3.13 British Health and Safety Executive (HSE)

This agency, the British equivalent of the American Occupational Safety and Health Administration (OSHA), has headquarters at Baynards House, 1 Chepstow Place, Westbourne Grove, London W2 4TF. It has responsibility for industrial safety in England, Scotland, and Wales. In Great Britain, construction industry safety problems are similar to those in America, but sometimes we use different approaches to their solution. In 1983, to heighten public as well as worker awareness of the importance of construction safety, a campaign was conducted to make that year the "Year of Construction Safety." A "Site Safe 83" logo, and the slogan "Keep Safety in Your Sites" were widely used on posters and stamped on letterheads by contractors. The problem of worker attitude toward safety was addressed in a letter from Buckingham Palace signed by His Royal Highness The Prince Philip, Duke of Edinburgh KGKT.

> Construction work is tough and the people who work in construction need to be tough, but there is no reason for them to be careless. Accidents can kill or permanently injure the tough, just as easily as anyone else.
>
> I hope the "Site Safe 83" campaign for sensible precautions against accidents will succeed in making everyone working on construction sites realize that it is possible to be both tough and careful at the same time.

The HSE distributed folders containing about two dozen brochures on construction hazards. Among these was a construction site basic safety checklist. (Appendix K includes some of the items from this checklist.)

HSE annual summary reports include articles about new safety developments and have included single-paragraph descriptions of accidents that were avoided as well as those that caused injuries or property damage. Anyone seeking a fresh viewpoint may find it helpful to read some of the British publications. Their approach is less authoritarian than ours. They seem to prefer to explain rather than to issue orders, and to issue general rather than specific instructions. HSE inspectors have authority to act, but I sense they like the Teddy Roosevelt policy of carrying a big stick while speaking softly.

2.4 Monitoring and Adaptation

Running a safety program is like educating a parade. Hardly a few days may go by before a new subcontractor, new workers, or new foremen show up. They know nothing about your safety activities unless you tell them what has been previously done. Perhaps new arrivals need only be given a brief summary statement. A tool box safety meeting the first morning on the site may be appropriate to explain the safety rules, distribute safety gear, and check tools brought onto the site by workers. If the project is such that unskilled workers without previous experience must be hired locally, it may be necessary to conduct more formal safety training sessions. The job of keeping the safety program alive never ends until the last key is turned over to the owner and the last worker leaves the job.

If there is a "near miss," investigate it fully and promptly. It may be a suitable topic for a meeting with people who could profit from your findings and thereby avoid a serious accident resulting from the same causes.

In the chapters that follow, topics are discussed in the approximate order in which the hazards first appear on a typical building construction site, but nobody can guarantee that the type of accident covered in one chapter may not occur before or after the stage of construction in which I have arbitrarily assumed it will be appropriate to include it. Safety engineers are therefore urged to consider their own projects and to try to anticipate trouble before it occurs. The timing of a program is important for maintaining interest. If there is a news story one night about a crane accident, and the safety engineer puts up some posters about crane safety before workers arrive on the job the next morning, they will certainly pay more attention to the posters than they would have had they not heard about the accident the night before.

A poster, slogan, or safety-first sign loses effectiveness very quickly. Who reads yesterday's newspapers? Notice the big signs that are often used that read "_____ days since the last lost time accident. Let's make it _____ days." Every day the numbers are changed (Figure 2.1). If paychecks or any other mailings to employees go out from the office, consider having a safety news sheet enclosed. It is more likely to be read at home than on the job, assuming it is newsworthy and well written.

Try to sense whether the safety program is moving along and creating interest, or is in the doldrums. Every program is apt to bog down occasionally and need a shot in the arm. Plan something ahead for those times. A slogan contest with tickets to some sporting event may be timely. If a considerable amount of work has been accomplished without a lost time accident, the local paper may run a story. Enlarged copies of such stories make good posters. The National Safety Council has a wealth of information and publications that can be used. Write them for one of their general materials catalogs if you are in charge of a safety program.

Imagination and a sense of humor help keep programs alive. When one company decided it was time to recognize the good safety record of a local group of workers, they and their wives were all entertained at dinner and their supervisor was given a box containing one lone Band Aid®, reputed to be all the supplies needed to refill his first aid kit. The dinner was enjoyed by all, there were no long speeches, and the award ceremony is still recalled by those who attended.

Families can be brought into safety programs. Sometimes a copy of a safety newsletter is sent monthly to every worker's home. It may deal with off-the-job safety as well as on-site safety. A small prize for the first dozen correct answers to a safety quiz or puzzle can stir up some interest. Possibly the winners may have their choice of any $10 or $20 item in a premium catalog. The amounts need not be large; I think it is better to give out several prizes than one large one, or to give people who have won more than once some special recognition.

The momentum of a good safety program carries over from job to job. Every time a project is completed without an accident it is proof that safety is possible, attainable, not a dream. When the same supervisors and workers appear at the site of the next job, they have the expectation that it will be run safely. The program is off to a running start.

Continuity is the cement that ties the other parts of a safety program together. Motivation starts it off, knowledge and awareness keep it on the track, but only continuity keeps it alive and moving. All three elements of the program are essential

Maintaining continuity and momentum sometimes requires adapting the program to changing conditions, and it is necessary to monitor safety programs in some way to be sure they are on the track. If the program is working reasonably well, the incidence rate over a period of a few months will not be very useful as an indicator of the efficacy of the safety activity. Keeping a record of near misses may be a better guide to the need to make changes. Something should be changed if only good luck has kept the

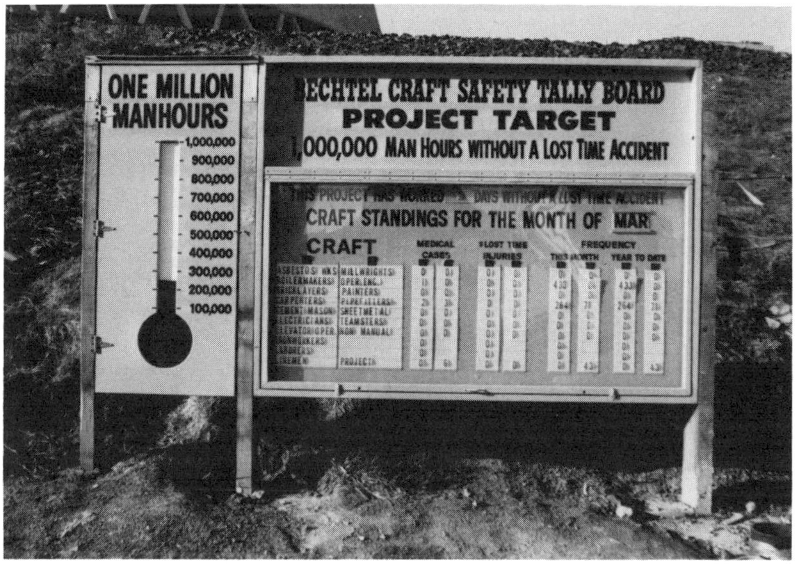

Figure 2.1 This photo of Bechtel's safety tally board at Philadelphia Electric Company's Limerick Generating Station was taken in April 1982. It showed a target of one million Manhours without a lost time accident. That goal was reached on December 31, 1982. The tally board shows safety data for each craft, including the number of medical cases, lost time injuries, and frequency rates, all kept up to date monthly. New workers are given a booklet on safe practices and are shown a videotape program on safety. Foremen get a 10-hour OSHA safety course, and they conduct weekly tool box meetings on topics selected by the project safety department. The Project Safety Supervisor's staff of 11 includes three nurses and can be supplemented by 34 additional people who are trained in fire fighting and first aid. Weekly safety meetings are attended by the Project Superintendent, his staff, craft stewarts, foremen, and other supervisors. The result: an incidence rate less than half the construction industry average. (Photo courtesy the Philadelphia Electric Co.)

incidence rate low. Watch the first aid cases; minor accidents could just as well have been more serious, but for chance.

Watch also for new situations that provide opportunities for accidents to happen. At the time a site was first surveyed, crane access may have presented no problem, but perhaps a new power line was installed while the project was under construction, and to get the cranes to the next project may require that they pass under the new power line. Another job may be started nearby; will it have any effect on the safety of the job for which you are responsible? Consider local traffic, possible effects on the labor supply, and perhaps some effect on the water supply available for drinking, construction work, or fire fighting. Take a look at the program as a whole once in a while. Was it well conceived? Is it working out as anticipated? Is it too late to make improvements?

3

Preconstruction Preparations

3.1 Applicable Regulations and Responsibility for Enforcement

After bidding a job, and signing a contract, let's hope there will be no real surprises regarding local codes or other regulations. In this category, the prudent contractor should have taken the trouble to check the specifications and code requirements, so that if for example, the fire rating of doors and walls in the two documents do not agree, the discrepancy has been noted and the more costly alternative has been used in bidding. Now, however, it is advisable to doublecheck the plans for code compliance. Are exits of adequate width shown on plans? Do the doors swing outward where required? When smoke and heat detectors are required, do the ones on which you bid meet local requirements, or must another make that has been listed in the "tested and approved" book of the local fire department be used? Watch for all these minor items; it is part of the loss prevention program to avoid having to tear out and replace safety items that do not meet the local code, and such codes are by no means uniform. Some even require compliance with an obsolete nationally recognized code, and compliance with the revised edition of the same code is not acceptable. Fortunately this is not common, but it does turn up, even in large cities where code revision is a cumbersome process and some inspectors delight in requiring compliance with the letter of the law. Give some thought to the possible personal reasons they may have; a word to the wise is sufficient.

Responsibility for enforcement of regulations may be different in different localities. Some cities have local codes that supplement or supersede state codes. Therefore, on one side of a boundary line on the map, state officials have the final word, but on the other side of the line, a local department has authority over the job. To avoid delays when it is time for work to be inspected, be sure you know who should be called. Inspectors are sometimes easier to get along with if they are consulted about a minor point or two before the work is done. If the code is open to interpretation and can be read in two ways, it helps to ask the inspector for his interpretation and then to do the work his way even if it costs a little more. Diplomacy is the art of compromise; concede the minor points and the major ones may not even be challenged.

Within your own organization, be sure workers know the lines of authority. If, for example, rules for the use of safety gear are set by the general contractor and apply to all subcontractors, let it be known whether the general contractor may stop unsafe work by the subcontractor's employees under ordinary and under emergency conditions. When construction work is being done in a manufacturing plant, and production work is also in progress, the plant foreman and the construction foreman must have clearly defined routine and emergency authority. Sometimes both agree that the other may stop any unsafe work or work that may be hazardous and then immediately notify the other. An example would be the need of a contractor to do some cutting and welding while the plant maintenance or production men had to do some painting in an adjacent area. Either foreman should stop the work if an explosion could occur, but rescheduling the work to avoid the hazard requires agreement between the foremen or their supervisors.

In anticipation of possible disciplinary problems, union representatives should be brought into discussions of the safety program. If a worker must be discharged for willful endangerment of others, for repeated violation of safety rules, or for any other related cause such as drug abuse, there should never be disagreement between the union officials and the contractor. Both have a common interest in the safety of everyone on the job. However, there may well be differences of opinion regarding proper disciplinary action, and the time to resolve such questions and to reach a common ground of understanding is early in the planning of the safety program. When the union and management work together in setting the rules, enforcing them is unlikely to be too difficult. But watch out for trouble when you try to fire a man for an infraction of a rule the union representative considers unnecessarily rigid. He could be right, you know. People who write rules do not always know much about the details of a worker's job, and foremen have been known to look for an excuse to get a man off the job. The time to work out problems is before they become confrontations; this takes patience, foresight, and skill, but it is worth the effort.

3.1.1 Gifts, Contributions, Gratuities, and Bribes

It is no secret that some contractors and some public officials have profited from the violation of codes and safety regulations. Before construction begins, the safety engineer should recognize the possibility that there may be an accident in which there is loss of life, many injuries, or extensive property damage as a result of deliberate violations that were permitted in return for something of value that the participants in the deal would not voluntarily disclose.

No sensible safety engineer is likely to deliberately endanger his career by initiating an attempt to bribe a public official, but he should recognize the possibility that the owner, the contractor for whom he is working, or the building inspector may make an illegal arrangement for their mutual benefit, and expect the safety engineer to acquiesce. If he protests, he may be offered a few dollars or be given the choice of going along with the deal or finding another job.

I assume that those directly involved in illegal transactions are aware of the possible consequences and consider the profits worth the risk. I also assume that experts in such matters are equally adept at being sure someone else will take the blame if an accident occurs. These assumptions are based on the difficulty investigators have in placing the blame for violations, even when it is reasonable to assume those in a position to profit from an illegal deal were aware of it.

Therefore, any safety engineer who would be the obvious target of investigators should be alert to the possibility that he may some day be trapped as a scapegoat. If in the preconstruction planning of a safety program he gets the feeling that there might be payoffs for safety violations, he needs to consider how he should protect himself, the workers for whose safety he is responsible, and future occupants of the building.

James E. Clyde, P.E., author of *Construction Inspection** had many years experience as a design engineer, inspector, and project manager. His advice to young inspectors is equally applicable to safety engineers:

> I believe any technically oriented person who is honest with himself and others and who finds satisfaction in creating things can become a good inspector. He must accept the fact that to be truly successful he must be as Caesar's wife "above reproach."
>
> No honest person can accept favors, however small, from a contractor on one day and rule against him the next and keep his conscience clear for very long.
>
> Over the years some contractors have offered me expensive gifts and favors, all of which were rejected promptly. Later most of these contractors submitted large claims for extra work or requested the relaxing of parts of the specifications. Each case was considered on its merits and if deemed unjustified, was rejected with no fear of claims of past favors.
>
> Articles appear from time to time on scandals and payoffs in the construction field. As in all businesses and professions, there are some undesirables who cheat and exchange favors with the contractors to make a fast buck. Sooner or later they are exposed and their careers ruined. Don't let it worry you. Keep your own nose clean and you won't have problems.

*A volume in this series of practical construction guides, p. 352.

3.2 Insurance

Expert completely impartial advice is hard to get in the insurance field. For those who can afford it, the services of certified insurance consultants are available. They have nothing to sell but their expertise; they write no insurance policies. But most contractors must seek out an insurance agent whose advice they can trust and, as anyone who has ever been a salesman knows, it is difficult to maintain a professional attitude, do what is best for the customer at all times, and still sell a product. Some agents represent only one company, which may or may not be the best in all the branches of insurance an individual contractor needs; therefore, doing business with several agents may be necessary.

Independent Insurance Agents represent more than one company and can often select a combination of policies to meet specific needs, assuming they are fully informed about the offerings of all the companies they represent. It is most important to keep options open, to feel free to do business with the firm that has the best policy. Maintaining that freedom of choice usually means avoiding doing business with relatives or close friends; it is hard to turn them down when someone else has a better proposition. Anyone who knows his business will respect this attitude; friends who think they are entitled to your business for personal reasons may be the best in the business, but when they are not, it may be necessary to risk losing a friend. Agents who are experts in life insurance, home owner's policies, and auto insurance often have difficulty maintaining sufficient familiarity with business policies, and so agents with whom you have been doing business in those fields may not be the ones to see about the coverage you need as a contractor. Check those that have been insuring other contractors and building owners. It is important that policies carried by owners and by contractors be coordinated, so nothing is omitted or duplicated.

If several firms seem capable, you may want to ask them all some questions that will test their ability and willingness to make technical points understandable to you. Your agent should be an expert in liability insurance. Ask him about hold harmless clauses and the details of policies that the specifications for a job or two require that you purchase. It pays to spend some time finding the right insurance agent because if an accident occurs the policy may be of little value unless it was expertly written to protect your interests as well as to meet the specifications.

Contractors should understand the basic idea of the principle of subrogation. Briefly, it involves the right of an insurance company to recover losses after paying a claim to the holder of an insurance policy. For example, after a fire the company may be obligated to compensate the owner but after paying the claim the insurer may have the right, under the principle of subrogation, to sue any negligent party who may have caused the fire. It is important, therefore, when a contract states that the owner will carry certain insurance, that the contractor know whether his own interests are protected. He may need to insist that he be one of the named insured in the owner's policy, or that a waiver of subrogation clause be included in it. Such riders have been found of questionable value in some cases, and should be reviewed by an insurance specialist or a lawyer representing the contractor.

In these troubled times, with foreign agents claiming responsibility for bomb damage to focus attention on their causes, check insurance policies to see whether there are exclusions that could allow the insurer to avoid paying a claim because it resulted from riot, civil insurrection, undeclared war (both here or abroad), or labor problems. If a project is in any way connected with a foreign country, check out the problems carefully. Recent protest marches against companies doing business with South Africa and bombings of embassy buildings are typical. Even peaceful protest marches can delay construction.

In property damage policies, check deductable clauses and premiums. Processing small claims may cost more than they are worth and, by choosing a higher deductable amount, the premium on the policy is reduced and the cost of submitting the claim is saved.

Avoid duplication of coverage. You cannot collect twice for the same loss and, when two insurance companies quarrel about which of them should pay a claim, who can tell how long it will be before either of them settles.

Watch the duration of policies; avoid intervals of time of even one day between the termination of one policy and the beginning of protection on another. If you may need to have workers on a job after the owner has occupied it, be sure your workers are adequately covered by whatever policies remain in effect. Similarly, if a job is delayed or is scheduled to begin at a time that cannot be predicted exactly, there is no point in paying for insurance until it is needed.

Insurance policies may be void unless the premium is paid before a stated date. When a policy is written, it is common for a computer to process a bill immediately. While this is going on, if a rider is added to the policy, another bill may go out and it is easy to assume the second one is merely a duplicate of the one already received. If both bills are not paid on time, the computer may invalidate the policy for nonpayment of premium. As we all know, once a computer gets things mixed up, no amount of human intervention is likely to do any good. To avoid the possibility of an accident occurring while a policy is being held up in this way, I suggest paying each bill when it is received and straightening out errors after all the dust has settled. Don't give the company a chance to avoid paying a claim on a technicality.

Architects, owners, and contract officers who represent them are usually

TABLE 3.1. Insurance Checklist

Liability and property Damage, including coverage for:
Worker's compensation, personal injury to individuals other than employees, and possible inclusion of owners of contracting firm.
Contingent liability, umbrella policies, hold harmless.
Property damage, including collapse of structure, boiler and machinery, rented equipment, completed operations, fire, smoke, and water damage (including leakage of fire protection sprinklers without fire).
Automobile insurance, including possible liability for property damage by employee-owned or rented vehicles.
Life, health, and accident (insurance for excutives of firm).
Theft, vandalism, malicious mischief, strike, and riot insurance.
Bonds required by specifications.
Marine insurance on components to be shipped from overseas.
Special hazards, such as earthquakes, floods, mud slides, volcanic action, sink holes, archaeological finds and unusual hazards peculiar to the site.

very diligent in following up contractors to be sure all certificates of insurance are received on time and are in proper order. They may be less careful in meeting their obligation to send copies of policies to contractors when the contract requires the owner to do so. Follow them up in time to have your insurance advisor check their policies carefully. Remember that insurance paid for by an owner is written by *his* agent, and the greatest care will be taken to protect *his* interests. It is up to you, as a contractor, to see to it that if you are entitled to coverage you are actually protected. Otherwise you may find yourself exposed to loss under a subrogation clause or an omission. If you are entitled to coverage, you should be one of the named insured. If you are not covered, you may need to take out additional insurance. Watch particularly the responsibility for materials at the site that have not been made part of the permanent structure. Be sure scaffolding and equipment you own or rent has been sufficiently protected against fire, theft, and vandalism.

When specialized knowledge is needed to complete a project, and only one or two members of your firm can provide this expertise, it may be advisable to take out insurance to protect the firm from loss if either illness or death prevents those key people from handling their assignments. Some firms prohibit an executive and his probable successor from traveling together; others have meeting rooms checked carefully by their safety specialists before allowing important executives to assemble. The loss of life at the Stouffer's Inn fire in Westchester County, New York in 1981 made it necessary to reorganize the management of the firms whose executives were among the 26 trapped in the blaze.

Don't rely on verbal assurances from subcontractors that they have insurance. Get a certificate of insurance from each of them and check the coverage against the insurance you are required to carry. This is particularly important in the case of worker's compensation.

I suggest that contractors sit down with their insurance advisors and work out a checklist that covers all the items that should be considered when policies are prepared. Different companies, all offering the same protection, write policies in slightly different ways. Exclusions may vary, and riders may be needed in some instances. I have attempted to prepare a checklist that includes most of the topics that should be considered, but I am aware of the probability that individual agents will offer policies that combine the risks in a variety of ways. Also, few contractors will be interested in all the items on the list but, if it provides the starting point for a productive discussion, the list in Table 3.1 will have served its purpose.

3.2.1 Worker's Compensation Insurance

If you are one of those contractors who regards worker's compensation insurance as one of those unavoidable expenses to be taken care of by getting any available policy to cover the people on your payroll, I suggest you take a closer look at the matter. You may be overlooking a chance to save as much as half the premium. On the other hand, you may be responsible for medical and other expenses for a worker you do not even think is one of your employees.

Don't expect any insurance agent to anticipate all your needs without your help; even the most experienced agent needs to know what the contract documents say and what special hazards there may be at the site. Compensation insurance laws may be confusing even to specialists. Some examples follow, but first let's discuss the history of compensation insurance.

Until about the turn of the century, if a worker was injured on the job, to collect any damages he had to sue his employer and prove negligence on the employer's part. The employer could avoid responsibility if he could show that the worker himself was negligent, the accident was caused by a fellow employee, or the cause of the accident was a job risk voluntarily assumed by the worker when he was hired. Legal costs and these formidable obstacles to securing a favorable verdict made it almost impossible for the average worker to confront his employer successfully in court.

In the United States, the first workmen's compensation laws were passed in 1911, and soon all states had them. They made the employer responsible for all accidents unless he could prove certain limited defenses were valid. The pendulum had swung from one extreme to the other.

The old system certainly needed to be reformed, but in strengthening the worker's position the new legislation reduced his personal obligation to accept responsibility for the consequences of his own actions. Even if he was entirely responsible for injury to another employee, the boss had to pay. Safety became of greater concern to employers, but not necessarily to workers.

An employer may be permitted to use a worker's willful and wanton disregard of safety rules as a basis for disqualifying him from benefits, but this rule, if it is still included in a state law, is seldom used.

There is also a trend away from scheduled lump sum payments for loss of a member of the body, or permanent partial disability, in favor of continuing weekly benefits.

There is widespread recognition of the need to improve the system, but I think that there is equally widespread disagreement on how to do it.

Every state has its own worker's compensation law, and every one is different from all the rest. Efforts to make benefits more uniform and insure that workers all have reasonably adequate coverage have been only moderately successful. Meanwhile, other complications have increased.

When work must be done near water, the Longshoremen's and Harbor Worker's Compensation Act may apply. This is a federal law, and benefits are higher than under most state laws. Consequently, injured workers try to obtain the best coverage. When two workers were injured on a job that extended into the Hudson River in New York City, they claimed the higher benefits but lost the case in federal court. The Supreme Court confirmed the lower court's action by refusing to review the case. But in another case, when workers were injured in a shipyard, a federal court ruled that they were covered by the Longshoremen's Act. Other laws have a bearing on the subject, and it is beyond the scope of this book to get into the details. Just get the best advice available if you have a job that is near water.

Workers often think that worker's compensation laws are their only recourse in case of accident, and employers are hardly inclined to tell them about other sources of compensation. But in many cases, workers have successfully sued building owners, professional engineers, architects, equipment manufacturers, and contractors other than their own employers. Lawyers usually handle such cases on a contingency basis, sharing in the court-approved benefits or whatever out of court settlement may be made.

For example, a utility company hired a contractor to install a pipeline and the contractor hired a subcontractor to apply a hot enamel coating. A passing motorist lost consciousness, knocked over a barrier, and caused hot liquid to be dumped on one of the subcontractor's workers. He was entitled to worker's compensation from his employer, but he also sued the general contractor, charging him with negligence in failing to guard the excavation properly. The court ruled in favor of the employee.

In one rather unusual case, a state agency hired a contractor to install a pipeline in a location where it was necessary for trucks to back down a hill in order to move excavated earth and fill. A worker was injured, and the courts awarded him $132,000 when he sued the state. Ordinarily, this would have been a simple worker's compensation case, with benefits paid by his employer's insurance company, but state officials had approved the contractor's work plan and the court ruled that the plan involved unusual hazards. Therefore, the state was responsible.

If a design error causes an accident an injured worker will probably sue the owner, architect, engineer, plan review board, any contractor other than his employer, or all of them. He is unlikely to sue a fellow worker; there is no point in trying to collect damages from anyone unless he has enough money or insurance to make it worth the bother. Third party suits, as such actions are commonly called, may be justified and are often won in court, but even when there is no basis for the suit the defendant's legal fees may be high. Cases may be settled out of court if trial costs might exceed the amount in dispute. All this legal activity causes many owners to require that all contractors carry hold harmless insurance to protect them and the design team from suits. If there is much money involved, the suits and countersuits can get out of hand. It may be practically impossible to determine what actually happened because everybody clams up immediately.

Perhaps the ends of justice require that we go through all this procedural routine, but I cannot help feeling that one original intent of worker's compensation legislation is being defeated. Certainly the idea that such laws would eliminate the need for court action and would assure speedy resolution of cases has been lost in the shuffle.

Lest my remarks about the possibility of large awards be encouragement to workers to seek more compensation than would be reasonable, I must report that when a court awards damages, any benefits already paid or due will be either deducted or adjusted by the court from the sum awarded. The system has its faults, but allowing anyone to collect twice for the same expenses is not one of them.

The worker's compensation laws and the publicity given to large awards when injured workers sue for damages have perhaps tended to make workers forget the consequences of accidents that *cannot* be covered by compensation laws.

A permanent disability award may bring a monthly check to a worker, but it cannot eliminate pain, take him on a canoe trip, or permit him to do a hundred and one things he would otherwise be able to enjoy. A young worker who might look forward to advancement or to owning his own company can lose these prospects as a result of an accident. If an ambitious worker had a second job or small business on the side, these sources of income may be cut off if he is disabled. Even when the total benefits from worker's compensation are added up, the total is usually considerably less than pre-injury income. Overtime is no longer available. Payment is limited to a percentage of straight time pay, with maximum limits on weekly payments in most states. Benefits to other members of the family may be lost if the employer carried health and accident insurance with family coverage for workers. For minor injuries, there may be a waiting period before any benefits are paid.

In safety meetings at which worker's compensation benefits are described, I have noticed that speakers are more apt to stress benefits than to point out losses to workers. This is only natural. Most speakers present the employer's viewpoint, and tell the workers what they are getting as a result of insurance. Workers might be more careful if they were more aware of what they may lose.

General contractors who subcontract all work and have no construction workers on their payroll may assume they have no need to carry worker's compensation insurance. This is only partially true; it is important to carry insurance on office and supervisory personnel, and to be sure all the subs and sub-subs carry compensation on all their workers. If any do not, the general contractor is considered the responsible employer in case an accident occurs. Without insurance he will probably be in violation of the state insurance laws. When total and permanent disability results from an accident, the uninsured employer can face financial ruin.

Some workers claim to be independent contractors; the term is used very loosely and many who use it are simply attempting to avoid paying social security and worker's compensation costs. Any contractor who hires subs should insist that they provide certificates of insurance, proving the types of coverage and limits of liability. Without such proof, the general contractor will find that when the time comes to settle accounts with his own insurance company he will be liable for worker's compensation insurance premiums for all the uninsured workers.

Many small subcontractors who do painting or incidental work are operating on a shoestring; they have old or inadequate equipment and may be so anxious to get work of any kind that they will take dangerous jobs nobody else wants. Any owner who hires an outfit of this kind without knowing for certain that the sub's insurance coverage is all in order is taking a big risk.

Some states permit independent contractors, partners in a business, or officers of corporations engaged in contracting work to include themselves in the worker's compensation payroll list. Even when executives or supervisors carry health insurance policies that include protection against loss of income in case of illness or accidental injury, those policies may exclude work-related accidents or illness. Needless to say, such exclusions make it possible for the insurance company to reduce the premium on the health insurance coverage and, unless the fine print is read carefully, executives may consider it cheaper to carry personal injury insurance than worker's compensation, totally overlooking the possibility they may need both. Insurance agents should be able to resolve questions of this kind, provided the problem is presented to them. At the same time that these questions are asked, be sure to identify the proper classification that applies to supervisors or executives. The percentage of time they spend at the job may determine whether they fall in a category with a high or low premium per $100 of salary. For high salaried executives, a difference of a few dollars in the premium rate can be an important item in the budget. By reducing the time spent at the job by a day or two per month, it may be possible to stay out of a high-rate category.

State compensation laws may make coverage of some employees optional but, if you choose to omit them from any compensation policy, be certain you know what responsibility you may have in case of an accident. State laws differ, so it is hard to say what should be done in any specific case without carefully reading the local law.

I do not want to get into too much detail regarding commonly-found differences between state worker's compensation laws, but perhaps a simple listing of a few such items will serve a useful purpose if it does nothing more than emphasize the need to know what laws require. The following are typical:

Injury or death caused by an accident while the worker was under the influence of alcohol or drugs may exempt the employer from responsibility. This exemption may be hard to substantiate, particularly if the accident is fatal.

Long-term disability benefits may be subject to automatic adjustment to compensate for inflation.

The Black Lung Act may apply rather than the state worker's compensation law.

Sometimes disability benefits paid under the federal social security law are deductable from worker's compensation payments.

If a worker has long-term partial disability but has returned to work, second injury fund benefits may be paid if he has another accident on the job.

Some states require that insurance be bought only from a state fund; others permit a variety of options.

Minors have special status in some states. Watch this carefully when the age of an applicant may be questionable. Similarly, investigate the status of part-time workers participating in vocational training activities.

Occasionally, only work that is considered "hazardous" need be covered by compensation insurance for the workers.

If death or disability results from occupational disease, there may be a long interval between the first and last exposure to the injurious substance. Meanwhile, the worker may have been employed by a number of contractors. As a general rule, I think the most recent employer is apt to be considered responsible, but I anticipate this point will be challenged and may be the subject of more specific rules.

The status of women in the work force has yet to be well defined;

relatively few are doing construction work, and as problems arise, revisions of laws are to be expected.

When state laws are changed, transcripts available for distribution may be out of date by several years. Check critical items with state officials before relying upon published guidelines.

3.2.1.1 SAVING MONEY ON WORKERS COMPENSATION INSURANCE. The base or "manual" rates used to compute worker's compensation insurance policy premiums differ in different states, often to a surprisingly large degree, but within each state all insurance companies use the same base rate table or manual. A typical rate manual includes several hundred job classifications, and for each classification there is a corresponding premium per $100 of straight time wages paid the worker. Thus, an employer of 10 workers for whom the base premium was $15.00 per $100 of payroll would be quoted a base premium of $15,000 on a payroll of $100,000 during the year. The base rate depends upon the loss rates, as analyzed by the state-approved rating agency, and for construction workers rates in excess of 25% of the payroll are not uncommon. It is therefore important to classify workers properly.

For example, carpenters may be in class 5645 if they work on detached 1 or 2 family dwellings, class 5651 if they work on dwellings three stories or less in height, class 5437 if they do only interior trim and cabinet work, and in class 5403 if their work is not described by any of the other classifications. For these four classifications, base rates range from under $6.00 to over $12.00 per $100 of straight time pay. If the paymaster does not indicate in which classification each worker's pay should be listed, the insurance company will charge the highest rate for any classification that applies to the trade.

Special rules apply when the same worker spends part of his time in activity that falls into two or more classifications. In Connecticut, the base rate for roofers is more than twice as high as the highest one for carpenters. Therefore, if a carpenter spends more than a certain percentage of his time doing roofing work, the premium applicable to a full-time roofer may be charged for all his wages. By going over the way payroll records are maintained with your insurance agent, it is possible to make considerable savings. He should be well informed about the possible classifications but only the employer can determine which of the alternative ones applies to each worker. Sometimes, by watching job assignments, it is possible to keep from exceeding the maximum permissible percentage of a worker's time spent in the high risk activity, thus avoiding having to pay a much higher premium on his total wages.

Insurance companies compete with each other by means of discounts below the base rates approved for the state in which the work is being done. This is permissible, provided all customers are treated fairly, without arbitrary favoritism. Shopping around for the best plan can reduce your premium payments by a considerable amount; failure to take care of this detail may give your competition the opportunity to bid enough lower to get closely bid jobs. With a good safety record, as much as half of the premium may be saved, and on some high risk jobs this can amount to 15% or more of the payroll.

Total base premium discounts are very common. The percentage discount off the base premium total depends upon the size of the premium, regardless of the amount paid out to settle claims. For small firms, the percentage discount is small, but it costs insurance companies little more to audit and bill large accounts; therefore the larger the account, the greater the base premium percentage discount will be. Total base premium discounts may seem unfair to small contractors, and they do give an advantage to larger firms, but this competitive rate adjustment is usually approved under state laws.

There are many plans that reduce compensation premiums when the contractor has a good safety record, and such discounts may be large enough to pay for the safety program with money to spare. One firm offers a retention plan type discount in Connecticut that reduces the premium if the safety record is good, but does not increase it if losses exceed the base premium. I have permission to summarize a typical proposal, but was requested to withhold the company's name.

In this proposal, the standard or base rate premium was estimated before the job began, and would be recalculated at the end of the job when payroll figures were available. In the proposal, the base premium was $45,268 on the basis of the estimated payroll. For this size account, the base premium discount would be $3799; therefore the maximum premium to be charged, regardless of the safety record, would be $45,268 less $3799, or $41,469. Even if losses exceeded $45,268, $41,469 would be the total to be paid. However, if losses paid by the insurance carrier were 50% of the base premium or less, a refund would be due, according to the schedule in Table 3.2.

The retention type plan just described is commonly offered to moderate size accounts, and has the advantage of guaranteeing them a maximum premium along with substantial discounts if the losses are less than half the base premium. Larger accounts may be offered greater discounts for better safety records, but without the guaranteed maximum premium. If losses exceed 100% of base premium, the insured and insurer share the higher losses. If a contractor has a good safety record and anticipates that losses will continue to be low, this form of contract, commonly called a retrospective rating plan, is attractive. Plans vary; often the net premium depends upon the average losses over a three-year period.

In some states, contractors may be self-insured, paying all their losses themselves and saving all insurance premiums. However, in anticipation of the possibility of catastrophic accidents, even the largest self-insured

TABLE 3.2 Example of Loss Retention Worker's Compensation Insurance Proposal*

Total base premium, as determined by payroll audit	$45,268
Base premium discount, approximately 8.4% of total base	(3,799)
Maximum net premium, regardless of losses paid	$41,469

Premium adjustment schedule, depending upon losses:

Actual Losses	Losses as % of Maximum Base Premium	Total Net Premium Cost	Saving to Contractor
$ 0	0%	$17,564 min.	$23,905
$4,527	10%	$22,996	$18,473
$9,054	20%	$28,429	$13,040
$13,580	30%	$33,860	$7,609
$18,107	40%	$39,292	$2,177
$22,634 or more	50% or more	$41,469 max.	$ 0

*Note that with this plan, the contractor's cost for worker's compensation may vary from a minimum of $17,564 to a maximum of $41,469, depending upon the benefits paid out to the insured workers, as tabulated in the "actual losses" column.

firms customarily carry some form of insurance. In order to process claims, such firms need an in-house insurance department, and discussion of such situations is beyond the scope of this book.

In states where insurance is compulsory and there is no state fund, contractors whose safety record has been bad may have to apply to the state in order to be put in the "assigned risk" category. The state then requires one of the insurance companies to issue a policy, often at the base rate without any discounts. Assigned risk accounts are distributed in some random or rotational fashion, so that no insurance company must handle more than the state considers a fair share of this undesirable business, and the insurer is not usually under any obligation to renew the policy if losses continue to be high.

A few states have state-operated compensation plans; private firms are excluded from competing with state funds, to the best of my knowledge, but laws keep changing and I hesitate to say with certainty what may be permissible from day to day. In Canada, provincial funds are popular; I understand it is the pattern followed in the whole country.

Before bidding on an out-of-state job, contractors should check the comparative premiums for worker's compensation. Some surprisingly large differences in base rates exist between even adjacent areas, and it is not even possible to say that one state's premiums are consistently higher or lower than another's. In Connecticut, for carpenters working on one or two family dwellings the base rate is over 70% *higher* than in adjoining New York state, but for general carpentry work, the rate in Connecticut is almost 30% *lower* than in New York. Anyone accustomed to working in Massachusetts, where base rates are comparatively low, might take a real beating if he hastily bid a job in Washington D.C. where some rates are over $75 per $100 of payroll.

If a compensation policy is written in one state and the job is in another state, workers who travel back and forth to the job will probably be covered by an extraterratorial clause in the policy, but it is customary to add an "all states rider." The company will then pay legally applicable claims, regardless of whether home state or job site benefits apply. It may be best to write the policy on the basis of the job location, particularly if most of the workers covered will be hired in that state. If the job happens to be in a state where insurance may be bought only from the state's own fund, the all states rider cannot be written.

In view of the considerable differences in benefits payable to injured workers in different states, it is important to them to know where they stand if an accident occurs. Some may not want to go to an out-of-state job if only lower benefits would be paid in case of injury. State laws differ in the percentage of weekly wages that will be paid to disabled workers, but they differ even more in the maximum and minimum dollar amounts that will be paid. If a worker is asked to follow the job, it is only fair and reasonable to let him know what effect it may have on him and his family. If, by going to the job, he will be eligible for higher benefits, the telling will be easy; he may not like what he hears if only lower benefits are allowed, but he will respect your integrity in telling him and would resent being kept in the dark if he got the information elsewhere.

If I seem to be avoiding making definite statements about this topic, it is because whatever I say might be wrong in some situations. It is necessary to dig into each case. Some will be easy, but don't count on it; allow time to resolve problems.

Most insurance companies that handle worker's compensation have some safety engineers who visit jobs and suggest ways to avoid accidents. Their recommendations are not binding; it is up to the contractor to act on the suggestions or ignore them. Occasionally, when the insurance inspector finds conditions very bad, he may recommend that the policy be cancelled, if the company has that option, or that it not be renewed when it expires. Companies differ greatly in the effort they make to help contractors avoid accidents. They are in a position to know a great deal about the cost of accidents and the most prevalent causes. Moreover, they have a direct financial incentive to reduce accidents; even when the contractor gets a discount in return for a good safety record, you can be sure the insurance company also shares in the benefits. Contractors who fail to ask for help from their insurance carriers are missing out on benefits for which they are paying as part of their premiums. Even when they must pay a nominal charge for literature suitable for distribution to workers or for other safety program aids, it is likely to be the best bargain available.

When competing companies quote comparable net premiums, I suggest the contractor take a close look at these "free" benefits. In a highly regulated industry, giving quality assistance to customers may be the best sales tool the company has, and I know of no industry in which this is more true than insurance.

Before leaving the topic of worker's compensation insurance, I would like to speculate on the possible influence of unions. The extent to which investment of union pension funds has made workers, in effect, part owners of businesses and creditors of even large cities suggests that the next development may be the sale of workmen's compensation insurance to employers. If an employer were to buy compensation insurance from the unions with which he had union shop or closed shop agreements, it would increase the union involvement in safety in several ways. All compensation payments would decrease union assets, so claims would probably be closely monitored for possible fraud. If an injured worker was collecting total and permanent disability payments but was no longer entitled to them, it is unlikely that other union members would rat on him to an insurance company but, if payments were from their own union treasury, they might have a different attitude. Also, in the process of studying claims, union insurance experts might detect more instances of third party negligence than would insurance company adjustors who were less intimately familiar with local job conditions. Furthermore, peer pressure to work safely would be likely to increase, particularly if workers on each project shared with the employer in premium refunds due when losses were less than the base rate premiums approved in the state. Finally, union-employer cooperation in drafting and enforcing safety rules would be enhanced; the union would have a direct interest in enforcement of rules its members helped to draw up. On the other hand, it might become a topic for negotiation when the cost effectiveness of a rule was debatable. All this would have to take place within the framework of OSHA regulations and state laws, but there are often enough alternative solutions to safety problems to make such negotiations possible. I would expect that sharing of savings that result from improved safety would be negotiable.

The insurance industry would certainly be affected; companies might become involved in several ways. Some might sell insurance to unions, covering large numbers of workers, and unions might then negotiate smaller contracts with employers. Private insurance companies might cover catastrophic losses only, as they do with large companies that are self-insured for most claims. Specialists from insurance companies would undoubtedly be hired by unions or perhaps companies could offer advisory services for a fee.

These developments may never occur, but when union funds are available for investment, and insurance companies find worker's compensation a profitable field, I think the possibility that some of these ideas will materialize and grow cannot be overlooked. If so, I anticipate that unions would try to make their proposals profitable to employers in order to gain entry into the field.

3.3 Anticipating Emergencies

Accidents are just one of the many kinds of emergencies that should be anticipated and for each of them someone should be responsible for planning action and taking charge when necessary. Time is of course the critical consideration; unless people know what they are expected to do it can be too late to act effectively. This is especially true of supervisors. The chain of command must be known, so the person who should be in charge will take over immediately. The project manager may not necessarily be the person to handle every type of emergency, particularly if the site is a large one and he may be hard to find or if he is handling several small projects at the same time.

3.3.1 Injuries

The highlights of 29CFR 1926.50 on medical services and first aid are:

1. The employer shall insure the availability of medical personnel for advice and consultation on matters of occupational health.
2. Provisions shall be made prior to commencement of the project for prompt medical attention in case of serious injury.
3. In the absence of a readily accessible physician or medical care facility a person having certified qualifications to provide first aid must be available on the job.
4. First aid supplies approved by the consulting physician must be easily accessible and checked at intervals of one week or less. Supplies must be in sealed packages and kept in a weatherproof container.
5. Appropriate means of transporting injured persons or for calling an ambulance are required.
6. Telephone numbers of physicians, hospitals, or ambulances must be conspicuously posted.

To comply with parts 1 and 2 it is necessary that the physician agree to assume responsibility for the care of workers. Just picking the name of a doctor out of the phone book is not enough. He may already have commitments to all the patients he can see. He might be ill himself, or semi-retired. He could also be specializing in one branch of medicine and prefer not to handle industrial accidents. Be sure he realizes he may be needed at any time and will keep his answering service informed of his whereabouts at all times. Alternatively, he may have a radio paging service. Nobody can be available 24 hours a day every day of the year and the physician who provides back-up service should be aware of agreements with the doctor. Figure 3.1 indicates common types of medical emergencies.

When making arrangements for emergency care at a hospital, be sure to investigate their admission requirements. They may insist that someone guarantee payment before they will provide service and, while the worker's employer must be willing and able to do this, there may be delays if the hospital's routine is not understood. Some hospitals have been unbelievably callous in this regard. In May 1982 a painter was cleaning brushes at his home when the solvent caught fire. He suffered burns on almost every part of his body. One might assume that the nearest burn center would take care of him first and worry about payment later, but he was not covered by insurance and no less than 30 different burn centers refused to admit him. The governor of Georgia finally guaranteed payment from the state emergency fund, and the man was admitted to a hospital in Baltimore in critical condition. Until a person has been hospitalized himself,

Sprains and Strains	XXXXXXXXXXXXXXXXXXXXXXXXXXX	27.3%
Cuts and Punctures	XXXXXXXXXXXXXX	14.4%
Fractures	XXXXXXXXXXXXXX	13.7%
Bruises	XXXXXXXX	8.2%
All Diseases	XXXX	3.6%
Abrasions	XXX	2.9%
Burns	X	1.4%
Amputations	X	0.5%
All other:		
Classifiable	XXXXXXXXXX	7.9%
Nonclassifiable	XXXXXXXXXXXXXXXXXXXX	20.1%

Source: Occupational Injuries and Illnesses in Connecticut, Connecticut Labor Department, 1978.

Figure 3.1 Common types of injuries and illnesses in the building construction industry.

it is hard to believe the amount of red tape that can be involved. Not all hospitals are hard to deal with, but enough of them are to make it important to investigate before rather than after an accident.

Even when there are good local medical services available, it is advisable to have someone with first aid training on the job at all times. Many people have been instructed in the armed forces, or have been trained as life guards or volunteer firemen. Check with the local authorities about training and certification requirements.

OSHA requires that the first aid kit be approved, inspected, and accessible. I have found that it can be so accessible that it is impossible to keep supplies from disappearing. Minor items such as Band Aids®, headache remedies, and anything else that could be useful at a worker's home are apt to walk away unless kept in a safe place. I suggest that only a few people, at least one of whom will always be available, have access to the kit. One of them should be responsible for replacing items that are used. This plan helps assure that even minor injuries will be reported. Sometimes it is a good plan to have a well-stocked kit for emergencies and to have a small one accessible, in which a few Band Aids®, antiseptic, and such minor items are kept. If only one or two of each are left in the kit at any time, people are less apt to make off with them than if they see a bountiful supply.

When ambulance services may be available from several organizations, get the doctor's advice on the one to call. Some specialize in emergency work and know how to handle accidents; others are better qualified to transfer patients from nursing homes to hospitals and may be out on such calls when they are needed in a hurry. For very isolated sites, helicopter service may be needed.

To comply with the OSHA rule that telephone numbers of doctors and medical facilities be prominently displayed, I suggest that several copies of the list, including driving directions, be tacked up near each phone. In an emergency the top copy can be torn off and taken along with him by the driver, saving him the time to read it and make notes before getting on the way. Someone can probably read the details to him en route.

Be sure to keep such lists up to date. In even a few months phone numbers can be changed, one-way street signs may be put up, or other changes may be needed. Nailing up some scratch paper near the phones will also help keep people from marking extraneous numbers on the emergency list; I have seen some lists that looked more like a sheet from someone's date book than a medical facility list.

3.3.2 Illnesses

Very few of us feel entirely well all of the time. It is often a question whether we are well enough to go to work, and I guess everyone has at one time or another regretted showing up on the job when he should have stayed at home. It is every supervisor's responsibility to be alert to the possibility that a worker may be unable to do his job safely and to take action to get him to a doctor to find out what is the matter.

Signs of illness include erratic performance, irritability, slow response to questions or unsteadiness. The better the supervisor knows the person's normal behavior, the easier it is to detect conditions that make it unsafe for him to remain at work.

It is very important that supervisors concentrate on symptoms rather than trying to guess what the cause may be. When an individual shows up late, particularly after a weekend, and seems to be suffering from a hangover, it is easy to guess what is wrong, but a doctor may disagree. If you discipline a person for being drunk on the job and he claims a few days later that your action was unwarranted, your opinion has no more value than his. An arbitrator would give his attention to a doctor's testimony, but the impact of whatever anyone else might say is questionable.

Tranquilizers, such as Valium®, and sleeping pills can affect performance on the job, sometimes hours after the patient thought the effects of the drug should have worn off. Antihistimines and drugs used to control motion sickness may contain additives to reduce drowsiness, but the effects of the additive may not last as long as the basic drug. Pain relievers and prescription drugs for various diseases can have unexpected side effects if taken with even a small amount of alcohol. Allergic reactions are also a possibility. Consequently, the diagnosis of alcoholism may or may not be warranted when the untrained observer thinks that it is present.

Even after it has been established that a worker is an alcoholic, current medical opinion tends strongly toward treating the condition as a disease over which the individual has lost control. Obviously, a worker under the influence of alcohol, or incapacitated during the time he is sobering up, cannot be permitted on the job, but rehabilitation is often possible. The National Council on Alcoholism can provide valuable information on how cases may be helped by local organizations and, in view of its estimate that the disease affects between 5 and 10% of all workers, including management, there is a great need for the work it is doing.

Getting a worker to a doctor can be difficult. He often fears that the result of the examination may be the loss of his job. Assuring him that your only concern is for his safety and the safety of others may help overcome his objections. Also, the cost of the examination may be a factor if he does not realize that his employer will pay for the visit. If all else fails, it may be necessary to lay it on the line: you are responsible for safety on the job, you do not consider him fit to work, and it is up to him. He can go to the doctor and find out what is wrong or quit. If he does neither, he will be discharged, period.

In anticipation of confrontations of this kind, and possible job actions that may result, be sure to keep records of all serious reprimands or warnings. Let the worker know that a record of his violations of safety rules or his inability to meet work standards is being maintained. You may want to give him a copy of the memo recording each incident. Keep such memos brief and factual; "so and so was observed working above projecting rebars without using a prescribed safety belt; he was reminded of previous warnings that this equipment was to be used; he was reprimanded for continuing to violate safety regulations. This incident occurred at 3:00 P.M." Sign and date the memo, and put it in a file where it can be readily found. You may want to add a notation that the worker was given a copy of the note, or add that the incident was observed by certain other workers, whose names were recorded, and who were using the prescribed gear. Sending a copy to the union representative may be another desirable precaution. If you have difficulty enforcing a rule possibly the union will help. Unions do want safe working conditions on the job and, if you are cooperating with them, some help may reasonably be expected in return. It is better to get these problem cases out in the open with the union than to wait until an accident occurs. Observe that the foregoing note did not mention the reason why the worker did not use the safety gear; it concentrated on the fact that a violation occurred, was brought to the individual's attention, and records who was present to observe the trouble. The same procedure applies to cases when a person appears to be unfit to work. Just state the facts, and never mind about what you think the reasons may be. If the worker became angry, abusive, or threatened violence, that is of course a recordable fact. Just be sure your action did not justify the outburst. Keep cool and let others do the shouting.

A final word before leaving this topic: do not cover up for anyone. It is no kindness to allow someone to stay on the job when he is apt to get hurt or cause an accident. If he is an alcoholic, it may just prolong the time before he gets help if you let him get away with it. Until he realizes that it is up to him to shape up or ship out, he may just continue to get into more and more difficulties.

3.3.3 Noise and Hearing Protection

Impairment of hearing as a result of noise on the job is a permissible basis for worker's compensation insurance claims in many states. Therefore, to avoid the possibility of unwarranted demands, before assigning a worker to a noisy job it is advisable to find out whether his hearing has already been impaired. Even young people, formerly thought to be unlikely to have hearing difficulties, are now apt to have enough hearing disability to create a safety hazard because they may not readily understand verbal instructions. Comprehension difficulty that is not apparent in a quiet environment can be dangerous in a noisy situation. The higher the background noise level, the harder it becomes to distinguish between words that sound somewhat alike.

Before proceeding with this discussion of hearing, I think it may be appropriate to review a few definitions, even though many readers may find it unnecessary. The frequency or pitch of a sound, measured in Hertz (Hz), or cycles per second and its loudness, measured in decibels (dB) both have important effects on hearing. The lower frequencies, from about 300 to 3500 Hz are essential for the transmission and understanding of speech. The higher frequencies, above about 1000 Hz, are more injurious than lower frequency noise of equal intensity. Therefore, OSHA measures loudness on a "dBA slow response" scale that emphasizes the higher frequencies and reduces the weighting of lower frequencies. Portable instruments are available to measure noise on the dBA scale, and sometimes that scale is also marked "OSHA." If the true sound intensities in each of the seven octave ranges included in the dBA scale are known, it is mathematically possible to calculate the single dBA value, but the procedure is time-consuming and I therefore recommend that the type of meter that reads directly in dBA be used if field measurements are needed.

The weighting of the dBA readings is also important when the ability of hearing protection devices to attenuate sounds in various frequency ranges are evaluated. Always consider the reduction factor of the device as it affects sound levels measured on the dBA scale rather than at specific frequencies; the effect on the dBA readings may be many dB less than the attenuation effect reported at any specific frequency. Be sure to check this point when selecting hearing protection equipment; ask for technical assistance from the manufacturer if in doubt about catalog data.

When two or more sources of noise are present their total effect is *not* the simple arithmetic sum of their dB or dBA intensities. Two identical machines increase the noise level only 3 dBA above the level caused by only one machine. Four pneumatic hammers, side by side, raise the noise level by 6 dBA above the reading when only one is in use. Estimating sound levels created by several noisy pieces of equipment is difficult or impossible in the field, and therefore direct metering is always preferable.

Meter readings should be made at typical work locations because noise levels decrease rapidly when the distance from the source increases. Doubling the distance of the worker from the source of a loud noise reduces the dBA to which he is exposed by about 6 dBA.

Workers should not be exposed to impulsive or impact sound levels over 140 dB peak sound pressure level and the maximum continuous dBA should not exceed 115. A table in 29CFR 1926.52(d)(1) gives the permissible duration of exposure to sound louder than 90 dBA during any eight hour workday. I have reproduced it as Table 3.3.

Each line in Table 3.3 is a separate daily limit and if, for example, a worker has a total daily exposure of four hours to 95 dBA noise, he has had his full permissible exposure for the day and should not work any part of the remaining four hours of the day at any job where the noise level is

TABLE 3.3 Permissible Noise Exposures

Duration per day, hours:	Sound Level dBA Slow Response
8	90
6	92
4	95
3	97
2	100
1½	102
1	105
½	110
¼ or less	115

over 90 dBA. He could, however, work less than 4 hours at 95 dBA and less than the permissible total hours at another sound level, provided the daily total did not exceed a limit calculated by a formula given in 29CFR 1926.52(d) (2):

$$F_6 = \frac{T_1}{L_1} + \frac{T_2}{L_2} + \ldots + \frac{T_n}{L_n}$$

where:

F_6 = The equivalent noise exposure factor.
T = The period of noise exposure at any essentially constant level.
L = The duration of the permissible noise exposure at the constant level (from Table 3.3)

If the value of F_6 exceeds unity (1) the exposure exceeds permissible levels.

This formula is useful when men can be rotated from job to job during the day, thereby avoiding overexposure of anyone. Personal dosimeter instruments are also available to totalize noise exposure in accordance with Table 3.3. They are small enough to be carried in the worker's pocket and eliminate the need for any calculations.

Before assuming that a noise problem can only be solved by hearing protection devices, consider the possibility that noisy machines may be tearing themselves apart by vibration; do they seem abnormally noisy? Also consider whether the one that makes the loudest noise could be muffled, isolated in a shed, or moved farther from the workers.

Loud noises are a health and safety hazard even at permissible dBA levels because they make some people nervous or irritable. They become jumpy, and can be easily startled. Very high noise levels can cause pain, fatigue, and nausea. As previously mentioned, voice communication can be disrupted by moderate hearing disability and it becomes increasingly difficult as noise levels increase. Some form of hearing protection may therefore be desirable even when OSHA restrictions on exposure can be satisfied without their use.

Cotton wadding is not an acceptable hearing protection, but soft, easily shaped fibrous or plastic materials are made for the purpose.

Ear plugs must be fitted by "competent persons" to be sure they will effectively reduce sound levels reaching the eardrum without causing injury or irritation. A number of standard designs are available in a range of sizes. Custom molded inserts can also be obtained. In general, ear plugs are effective in attenuating both high and low frequencies, and consequently they make it harder to understand verbal instructions or warnings. Workers often resist using them unless they are well fitted and obviously needed.

Muffs allow more of the low frequency sounds to reach the ears, are generally more comfortable to wear, and easier to put on or take off than ear plugs. On the other hand, in hot weather, they may not be as acceptable as on cold days. The head bands may be worn over, or behind the head or under the chin, but in any case it is important that the muffs fit snugly around the ears. Flexible connections at the points where the muffs are connected to the band are useful.

The maximum degree of protection can be achieved by wearing muffs over ear plugs, but this is seldom necessary. Try to get sufficient protection, with as little discomfort as possible. The relative effectiveness of alternative products may be judged by the NRR rating in the manufacturer's literature; the higher the number, the greater the attenuation. Typical devices have NRR ratings of 15 to over 20 dBA. I recommend that job conditions be discussed with the manufacturer or some other knowledgable person before a final selection is made. If appropriate protection cannot easily be obtained, a second look at job conditions may be needed, including the possibility that workers may be rotated between several jobs or work areas, relieving those originally assigned to the worst areas from the necessity of working a full day there.

3.3.4 Fires

In planning the safety program, consider two classes of fires: accidents and arson. Preventive action and preparedness for both are needed.

Preventive action for accidental fires will be covered elsewhere in the appropriate chapters. See the index under topics such as welding, flammable liquids, and explosive vapors.

To avoid arson, be alert for situations that may provide incentives. If the conditions that led the owner to start the project have changed and he can see only continuing losses from its completion, he may be tempted to hire an arsonist. Arson for profit is difficult to prove, but it is less difficult to observe conditions under which the profit motive could be present. Watch for suspicious actions, such as the removal of valuable items from a remodeling job without an apparent reason. It may be advisable to arrange for a highly visible inspection by the insurance company or fire department, just to let the would-be instigator know that the possibility of fire is being carefully watched. If the insurance inspector finds the coverage excessive, he will smell a rat immediately. If he insists on only realistic valuation, the incentive to have the job burned will disappear.

Arson for revenge is another real problem. Watch out for the reaction of anyone who has been discharged for petty theft or whose reaction when leaving indicated the possibility of mental instability. If there have been suspicious fires at other jobs, have a talk with the head of the local fire department's arson squad.

When there is a possibility that vagrants may get onto the site at night or when work has been delayed or suspended, there is always the potential for trouble. Keeping them out may be difficult, particularly in cold weather, but it has to be done.

In addition to planning fire prevention measures, do not overlook the need for putting out fires if they do break out. Discuss possible problems with the local fire chief. Find out whether his equipment could handle a fire on the top story of the building you are erecting. Try to get the building's fire standpipe system into operating condition as early in the schedule as possible. If there is no local fire department, it will be necessary to organize your own fire brigade, and of course this takes some time. Be sure to fit it into the schedule early enough. Coordinate the medical, first aid, and fire protection plans.

You may want to read OSHA Safety and Health Regulations, subpart F of 29 CFR 1926 paragraphs 1926.150 through 1926.155 regarding fire protection at this stage of program planning in order to become generally familiar with OSHA rules. More detailed references to these regulations have been made in later chapters in connection with specific hazards.

3.3.5 Sabotage

The precautions against arson for revenge will also apply to the prevention of sabotage. However, if the project has any connection with the military establishment or with an organization that a special interest group may want to injure, sabotage may be a job hazard that must be given serious consideration. Consult the owner about this. He is in the best position to know how serious the problem may be, and may have had previous experience in coping with it. This topic is not one on which I feel it necessary to dwell at length, because it is most apt to be associated with special situations in which the possibility of trouble is apparent. I therefore am mentioning it only as a reminder that the possibility of trouble should be considered, discussed when it is apt to be a problem, and handled in accordance with appropriate police or security force activity.

Don't let a safety meeting become a school for saboteurs. Some topics should not be discussed in general safety meetings; reserve them for meetings with the workers who are directly affected only. For example, gasoline in a salamander is dangerous. The hazard must be controlled, but only those who take care of that equipment need participate in the discussion. If there is no need for gasoline on the job while salamanders are in use, simply removing all the gasoline cans could remove the hazard. Making a topic of this type the subject of part of a general safety meeting could give someone with a grudge the idea of how to cause a fire when he wasn't even near the job.

Listen to the questions people ask in meetings. Are they related to the safety aspects of their own jobs, or are they fishing for information for which they have no legitimate need. For example, a question relating to the way to turn off the water if a sprinkler head is accidentally broken off may be a proper one if asked by a foreman in charge of work that might suffer water damage. An appropriate response would be that you will meet with him and show him what to do. On the other hand, the same question from someone having no real need for the information should be cause for suspicions, and certainly it would be poor judgment to tell everyone how to shut off the water to the sprinkler system. Have you noticed that it is often reported that the sprinkler system was out of order when a fire of suspicious origin is investigated?

If someone accidentally sets off a security system alarm, don't be too quick to turn it off. By running directly to the disconnect switch you may be giving someone exactly the information he was after. Plan ahead, so you know how you want to handle this type of situation. During the construction period you may want to have a temporary disconnect wired into the system so that the details of the final system can be kept secret.

3.3.6 Theft, Robbery, and Vandalism

In some localities, thieves will make off with anything that they can sell, even if it has only scrap value. They often operate quite openly, even during normal working hours, and it is dangerous to interfere. Therefore, it is a concern of the safety engineer whenever things disappear from the job. Fire extinguishers may be stolen, and even a whole truckload of material may be driven away while the driver who was making the delivery is getting a signature on the shipping papers.

When workers are paid in cash, their personal safety and the safety of everybody who handles the payroll can be endangered by criminals. On one job, I recall losing a friend, a subcontractor. He paid his men in cash and was shot as he was going from one location to another. The crime was never solved.

Illegal drugs are so expensive that most addicts resort to some kind of criminal activity to get money. Consequently, where addiction is prevalent, there are many muggings and losses of personal property as well as construction materials unless adequate control methods are used. An associate had trouble with his car en route to work, and left the hood up while he went for help a block away. He found his battery and carburetor gone in less than 15 minutes. On a remodeling job, it was necessary to rip up and relay the asphalt tile in a large room because part of the material was stolen while work was in progress and replacement material did not exactly match the original shipment's color. In similar parts of town, copper tubing and valves will disappear overnight unless there is a watchman or guard dog on duty.

Police protection is often inadequate in such neighborhoods; there have been many instances where police have been unable to control looting after a break-in and, while I have not personally encountered it, friends have noted that they did not think local policemen were reluctant to pick up a few things when they thought the insurance company would pay for whatever was stolen during a burglary.

Vandalism that does not profit the perpetrators but is costly to the owner is so common as to warrant no comment.

Recognition of problems such as the foregoing at an early stage of the job is essential if they are to be controlled. The safety engineer who anticipates trouble and takes steps to prevent it has a better chance of avoiding difficulties than the one who waits until he is forced to recognize that he has a problem. In general, I think most thieves are lazy; if it takes too much trouble to steal from one job, they will go to another.

Keys and master keys need to be controlled. It is much more convenient to have one master key than to have to bother with 50 different ones, but suppose the master key gets into the wrong hands. Would it be necessary to rekey every lock on the job? Should some doors have two locks, so two different keys are needed to open them? For example, who has access to the powder magazine, and the key cabinet?

If petty theft begins to get out of hand and a control system is set up, consider having two systems. Make one quite visible; it will discourage the people who will take something if there is no risk involved, but who will stay honest if they think they may be caught. Keep the second security system quiet. Set it up to catch the clever ones who think they can outsmart the other system. For example, if standard packages of bolts and nuts are disappearing, a lunch box inspection at quitting time may be a visible checking method. Nobody will probably be caught after the first day, but thefts may drop. At the same time, dust the next shipment with fluorescent powder and put it aside where it should be safe. If it disappears, a simple portable black light may be all that is needed to find out where the boxes are going.

On large jobs, it may pay to order items subject to pilferage with some special feature that makes it possible to identify them if they are resold, or to make them useless if taken home. Some subway systems use light bulbs with left-hand threads. Antifreeze can be colored by adding a dye at the supplier's plant. The supplier may be willing to label it with a misleading warning that it is intended for use in construction equipment only and that it may damage automobile engine parts. When there is a choice between equipment that will work on two voltages and you have both available on the job, choose the one least likely to be usable elsewhere. For example, tools may be powered by single phase or by three phase motors. Which do you think would be most apt to be stolen?

3.3.7 Earthquakes

Many municipal codes require that designers anticipate the possibility of damage by earthquakes, but enforcement generally leaves much to be desired. Most people discount the risk, but others take it very seriously. The possibility of major earthquakes was the topic of discussion at a meeting of over 300 earth scientists and engineers in 1981, under the sponsorship of the Building Seismic Safety Council, the Earthquake Engineering Research Institute, the Seismological Society of America, and the U.S. Department of Energy. Their discussions indicated that there is a strong possibility of major earthquakes in the central and eastern parts of the United States before the year 2000, in addition to the widely publicized risk of damaging quakes along the San Andreas and other faults in California. Alaska is of course a well-known area where disturbances can be expected at any time.

Most engineers will recall that the Richter scale is used to indicate the intensity of earthquakes, and that it is logarithmic. A shock of unit intensity is about equivalent to dropping a 4000 pound weight about 12 feet anywhere on the site. Minor tremors, under about 2 on the scale, are not usually noticed but are readily detected by seismographs. For every increase of one number on the scale there is a 10-fold increase in earth motion and about a 32-fold increase in the energy released. Consequently, a quake of 7 or 8 on the Richter scale releases millions of times more energy than one of unit intensity and is capable of leveling structures many miles from its epicenter.

When considering geological events, one's usual way of thinking about time must be adjusted. It was only 10,000 years ago, a mere moment in geological time, when there was ice a mile thick over all of Canada and most of the northern states. Yet we tend to think that a volcano is extinct if it has been 100 years since it erupted. We should actually consider that an area where an earthquake occurred one or two hundred years ago as an active zone. It has been less than 100 years since the famous earthquake leveled San Francisco in 1906. People are concerned about the San Andreas fault in California, but relatively little has been written about dangers elsewhere in the continental United States. One epicenter of quakes is near the confluence of the Ohio and Mississippi rivers, close to New Madrid, Missouri. Recent studies there show that quakes with magnitudes of one or more are occurring about 150 times per year, but only about one in 20 is noticed by the local inhabitants. Still, the potential for a major quake remains. In 1811–1812 three quakes with probable magnitudes of 8.4 to 8.7 occurred there, and the tremors spread over a wide area. Church bells rang and pavements cracked as far away as Washington D.C. In 1976, the epicenter of a magnitude 5 quake was at Marked Tree, Arkansas, not far from New Madrid, and the shock waves were felt in seven states.

In June 1982 there were four earthquakes with intensities from 2 to 3 near Moodus, Connecticut but such occurrences are expected by the local residents. The Indians who were there before the colonists attributed the local noises of earth tremors to battles between good and evil witches in deep caverns where the town's namesake, Machemoodus, ruled from his sapphire throne. In 1638, Connecticut's Governor Winthrop was 15 miles from Moodus when he noted that a quake shook the ships in the harbor and the earth was unquiet for 20 days. In 1791 one strong quake was followed by over 100 aftershocks that felled chimneys and opened crevices in the earth. That quake was felt as far away as Boston and New York. From that time until 1979, at least 90 more quakes were reported and, since then, when a modern seismograph was installed, over 100 small tremors per month are being detected.

Seismologists have noted that for any given location, the number of quakes is roughly inversely proportional to their intensity. Just as there is a tenfold increase of intensity for every increase of one number on the Richter scale, there will probably be a tenfold increase in the number of earthquakes of one less degree of intensity between the more severe quakes. In other words, if there has been a quake of intensity 4 in a given area on the average of once in 15 years, there will probably have been 10 of intensity 3, 100 of intensity 2, and 1000 of intensity 1 in that vicinity during the same time period. Consequently, contractors should be aware that where major quakes have occurred, smaller shocks should be anticipated much more often.

These shocks, even though they are unlikely to be noted by workers, are the equivalent of driving piles with a heavy hammer anywhere on the site. They would certainly be observed if repeated in a rhythmic pattern as would be the case of a piledriver but, while a structure is in an unstable condition or concrete has just begun to set, I would hesitate to predict the consequences of one or two earth tremors. Where tremors are frequent, they may conceivably have a long-term effect on concrete, glass, or highly stressed metals. Therefore, earth movements deserve serious consideration, and certainly when a proposed structure is complex and innovative, it would be advisable to consult the state geologist before proceeding.

3.4 Community Relations

In large cities where construction jobs are a common source of noise, congestion, and dust, people become accustomed to the annoyance. They come to accept it as a necessary evil, a part of the price to be paid for living in the city.

In a small town the impact of a big construction job can be devastating. Construction workers may outnumber the local residents. Outsiders are apt the be treated with suspicion or resentment and often these feelings are well justified. Workers with money to spend, time on their hands, away from families and friends, and stuck in what they consider a hick town can't be expected to stay out of trouble.

If you are the safety engineer for the project, your official duties may not include any interest in what workers do with their spare time, but if they land in jail or in the hospital the effect on the job schedule is the same as though an accident happened on the site. Consequently, you may want to include some public relations activity in your schedule. You may want to meet with local officials, men's clubs, women's organizations, and social groups to get their views and to explain to them the long term benefits of the project. When you see the problems from their viewpoints, and have their suggestions for minimizing problems, you can plan your program accordingly.

Publicity may be an important part of the plan; it may involve taking a few ads in the local papers, providing speakers for meetings, or arranging for influential people to meet with the owner, architect, or designer of the project. The value of the project may have to be sold to the local residents in some tangible ways.

Obviously, some money will be spent there. It may benefit the local industries and merchants, or it may be disruptive. You may be making it difficult for small firms to keep their employees. You may be inflating prices, and perhaps making it impossible for residents to get their usual merchandise or services at the prices they have been paying.

Transportation to nearby towns may need to be improved to increase opportunities for recreational activities and ways for workers to let off steam without stirring up trouble.

You may find that construction workers have more in common with local residents than either group expects. Find out what local fraternal or social organizations would be willing to issue temporary memberships to workers who belong to similar clubs elsewhere. The list may be surprisingly long. Within a reasonable driving radius you may find golf clubs, a rifle range, skeet club, ball park, a Veterans of Foreign Wars meeting hall, a

Masonic lodge, swimming clubs, and any number of other social groups. There are over 300,000 ham radio operators in the country; many of them enjoy sending messages for other people who are away from home. It would be surprising if none of your workers were not hams themselves.

You will probably be using heavy equipment that has never been seen in the area before. Could you spare a crane for an hour to do some job the town fathers have wanted to have done for a long time but could not afford. Maybe the little league needs to have its field graded or a boulder blasted out of center field. Vocational guidance in small towns is often lacking. How many youngsters have ever had a chance to talk face to face with a civil engineer or anyone else who had actual experience in the construction industry? Doing small favors such as these can have a real influence on how the community feels about what is going on.

As the job progresses and different specialists come to town, find out about their hobbies. You may find an amateur magician, some rock hounds, any number of good photographers with movies or slides available to them from home, as well as people who have had interesting experiences in foreign lands. It costs little to get them to speak to a group, and can go a long way to break down the "them" and "us" differences. You may even have enough in the discretionary petty cash drawer to buy some refreshments for such meetings.

I am rambling on a bit about such activities because at some stage of your project you may badly need some local friends. A worker may get out of line and whether you have an angry mob to handle or the incident is readily controlled by the local police depends upon the general attitude of the townspeople toward your organization. If you are sitting on a powder keg, it takes only one incident to start a riot in which workers and townspeople may both be injured and considerable property damage may occur. When the community has a reasonably cordial attitude toward your workers, there is little chance that serious trouble will suddenly errupt.

In spite of all precautions, it is of course possible that an accident may occur under circumstances that do not reflect favorably on your organization. I am talking about the kind of story you would rather avoid telling. Don't attempt a cover-up. The more you lie, the worse it will get. When something newsworthy happens, reporters need to write stories; try to give them facts, such as the number of injuries, that a structural failure occurred, or whatever basic information is going to have to be revealed sooner or later, regardless of who was to blame. Avoid the controversial or speculative parts of the story; simply say that it would be unfair to the people involved to discuss those items until all the investigations have been completed. Telling the story straight eliminates the possibility that contradictory reports will be issued, initiating speculation and rumors that are almost certain to be more damaging than the truth. Try to get the story out in such a manner that it does not generate unnecessary speculation or the suspicion that someone is lying. In general, the story is most newsworthy immediately after it becomes public knowledge, and weeks or months later, after the investigations have been completed, many of the unfavorable comments that might have been made in the first reports will create less of a stir.

Sometimes a project is controversial; people have strong opinions for or against its existence. They may not want to see a certain industry expanded, or the owner may be involved in international dealings that are opposed by certain segments of the population. If protesters suddenly show up to stage a demonstration, they may include people with sincere convictions but there is also the possibility they may be led by professional rabble-rousers whose true motives are simply disruption of the economy or the interests of whoever is paying them. In either case, the safety of workers can be affected, and the job schedule can be seriously delayed. Mob leaders may be experts, specialists in the art of exploiting prejudices and creating incidents. Owners and project managers need equally expert help in such situations.

I believe that demonstrators are primarily interested in publicity. They want the TV cameras and news photographers to portray the owners and contractors as the enemies of the public, exploiters, or despoilers. They often hope to provoke someone to lose his temper and commit some rash act. The shouting and arm waving is most apt to occur when the cameras are in use. I recall one incident when a prominent public figure was passing through the streets in a motorcade. About two blocks away, a few dozen students created a disturbance and were dispersed by the police. The whole incident was trivial but, when I saw the evening news, it was blown up into a major story, as though there had been an assasination attempt. I think the story was deliberately distorted to favor the editorial policy of the news agency during a political controversy. It certainly gave the impression that the official was unpopular, whereas the cheering crowds all along the route of the motorcade indicated that public sentiment was on the side of the visitor.

I admire the way some politicians handle abrasive reporters. They give calm, reasonable, noncommital answers to questions that could anger a less experienced person and cause him to make a rash statement that could be quoted out of context and thereby distorted. In a confrontation between protesters and a representative of a project owner or contractor, it is not necessarily the person who knows the most about the job who is best able to handle the situation. A spokesman who projects a friendly image, remains calm under stress, and is not easily provoked will probably do a better job than someone who is better informed but less experienced in handling hecklers.

If the lid can be kept on a potentially dangerous confrontation for a few days, time is gained during which tempers can cool and sensationalism becomes more difficult. The first day of a demonstration is more newsworthy than the tenth day; by that time the public has heard all about it and has gotten just a little bored with hearing the same thing repeated. Eventually, a more exciting story will claim the attention of the news reporters and without a source of news coverage, there is little point of demonstrating.

3.5 Personal Safety Equipment

The National Institute for Occupational Safety and Health (NIOSH) provides a testing, approval, and certification program assuring commercial availability of safe personal protective devices and reliable industrial hazard measuring instruments. NIOSH develops improved performance regulations, tests, and certifies (or approves) devices, and purchases approved products in the open market to verify the quality of manufacture.

The Testing and Certification Branch of NIOSH, 944 Chestnut Ridge Road, Morgantown, West Virginia 60505, is prepared to supply additional information about products. Lists of certified equipment can be obtained from the Division of Standards Development and Technology Transfer, NIOSH, Robert A. Taft Laboratories, 4676 Columbia Parkway, Cincinnati, Ohio 45226. Send along a self-addressed mailing label.

Lists include the certification number, commercial identifications of the products, some information about the applications for which products are suitable, and the addresses of the firms to which the approvals were issued.

Most libraries have a Thomas Register. It currently consists of eight volumes in which products and services are listed alphabetically with subheadings that indicate the locations of the principal manufacturers and vendors in each state and city. When the product listing includes a distinctive "Tomcat" logo, the manufacturer or distributor has published catalog information in one of the six product catalog volumes of the Register. The publisher's address is One Penn Plaza, New York, NY 10119. I suggest that safety engineers get catalogs from one or two good distributors who handle a wide variety of safety items. When new products become available they are usually featured by such firms. The principal disadvantage of dealing exclusively with distributors is their reluctance to list any manufacturer's complete line. Consequently, I have also found it desirable to get catalogs directly from manufacturers for major items. If the exact item is not handled by a local distributor, it can be ordered direct but, without the manufacturer's complete catalog, it is easy to overlook the existence of the most suitable item. The manufacturers are also best qualified to help users select a product to meet field conditions.

I also recommend that safety supervisors involve workers in the selection process. It makes it easier to get them to use the gear if they helped select it and if they know why it is needed. Often several items will be technically acceptable, but workers will find that one is more comfortable than the others.

Familiarity with safety gear is important when unexpected hazards are encountered. For example, when excavation work is in progress, chemical wastes may be uncovered. Fast action may be needed to avoid holding up the job and time can be saved if contacts have already been made with suppliers of protective clothing. Even when the problem seems to be unique, people who are in the business of making or distributing safety gear for many industries may have run into the same difficulty elsewhere and have exactly what is needed.

Construction work does not necessarily involve handling toxic or hazardous materials, but a considerable variety of such substances may be encountered. I have commented on some of them in Appendix C but before safety equipment is ordered I suggest confirming the choice by a call to the manufacturer of the product or the safety equipment.

3.5.1 Hard Hats

The American National Standards Institute Standard Z89.1-1981 recognizes four classes of hard hats:

Class A — General Service
Class B — High Voltage Resistant
Class C — Metallic Headgear
Class D — Helmets for Firefighters

For most construction work, Class A hats with a full brim or caps with a beak and only a narrow brim on the sides and back are suitable for protecting workers from falling objects or bumps. Be sure the headgear meets ANSI standards; some manufacturers sell lightweight bump caps that are not acceptable. Pay particular attention to the suspension. It must hold the hat above the head to cushion impact and should keep it from falling off readily. It should also be easy to adjust, clean, and replace. Liners should be available for cold weather. Some manufacturers offer disposable liners so clean hats can be issued to visitors without having to sterilize the whole assembly after each use.

When choosing general purpose hats consider the ease with which transparent face shields, hearing protective devices, and work lights may be attached. Many are designed to be worn by welders also. Most hats are the same size, but some manufacturers also offer small and extralarge sizes.

Class A hats vary considerably in weight; some are lightweight molded polyethylene, but heavy duty hats may consist of multiple layers of textiles and wire screen bonded together for maximum impact resistance. All approved Class A hats are therefore not equally suitable for all jobs.

Some Class A hats also meet the higher voltage resistance requirements for Class B headgear.

Metallic hats, Class C, must not be worn where there is danger of contact with live wires, but their light weight and reflective surface make them popular for hot weather wear.

Class D helmets have special features demanded by firefighters. Most have a wide brim front and back, a means for attaching an identification shield easily, and good resistance to high temperatures.

Hair styles change so quickly that long hair may be a safety hazard one year and not the next. However, good headgear should include a net or other restraint for long hair whenever it is in fashion and workers are near rotating equipment or sparks.

3.5.2 Eye and Face Protection

Workers who wear prescription glasses and, in fact, all workers except welders are very likely to be difficult to convince that they should wear goggles. Even when they get something in an eye, they often go right back to the same job without using their goggles. Compared to the loss of eyesight, the reasons they give for avoiding the goggles are ridiculously petty, but complaints of discomfort, fogging, and restricted side vision cannot be ignored. Eye protection must be "sold" to workers.

Fogging can be avoided in several ways. When eyes need to be protected against coarse particles such as chips, goggles with a wire screen instead of a plastic lens can be used. When an impervious lens is needed, the fog-resistant types should be used. In cold weather, a double lens will act like a storm window and help prevent condensation on the inside of the inner lens. Frames should have good ventilation openings. Some designs have screened inserts, others have many small perforations, and when liquids are to be handled there are splash-proof vents, some of which have permeable foam inserts. Under some conditions, fogging is not entirely preventable, but more can be done about it than is often assumed to be possible.

Sometimes workers complain that their goggles are fogged when the lens is actually only dirty, scratched, or pitted. When these troubles are corrected it should be pointed out that the goggles served their purpose; eyes were protected from whatever damaged the goggles.

Restricted side vision can be a valid complaint, but if glare from nearby welding is strong, some restriction of side vision may be unavoidable. Goggles with a tinted frame may serve if the glare is not so strong as to require cup type goggles or very dark frames. Also, welders may be able to work behind a movable shield, so that the glare problem is eliminated for other workers.

In the absence of the need for protection against glare, wide vision frames with a variety of venting arrangements may be considered. Even the lightweight "visitors" glasses may be sufficient for some workers. They usually have clear side shields and may have frames that fit against the face for the entire width of the frame.

Curved face shields that are open at the bottom and are made to be attached to a hard hat are often suitable. They give good protection with a minimum of fogging but of course they are useless if the particles in the air are being blown up under the shield.

When a worker's complaint is discomfort, it may be necessary for him to realize that his job requires that he put up with a certain amount of it in order to avoid severe eye injuries. On the other hand, goggles should be ordered that fit the worker's face, straps should be properly adjusted, and the venting should allow the maximum amount of air circulation consistent with the job requirements.

I have sometimes felt like a visitor from outer space when it was necessary to wear a hard hat, goggles, and a respirator all at the same time. I would have found it more comfortable to wear a full face respirator (see Section 3.5.3).

Welders are usually well informed about eye protection, but if they have to work in a confined space or are exposed to fumes created by welding or burning coated metal they may not recognize the additional dangers (see Section 7.6). Other workers who are exposed to glare from welding may need to be supplied with lightly tinted goggles with side shields.

Welder's goggles should be made in accordance with the requirements of American National Standard Z 87.1-1968. OSHA regulations are in 29CFR 1926.102, from which Table E-2 has been reproduced in Appendix D. Shades more dense than those listed in the table may be used to suit the individual's needs.

3.5.3 Respirators

Many contractors have no need for respirators from one year's end to the next. Therefore, when they do need them they may not recognize the critical importance of using the right kind in exactly the right way. Proper selection begins with recognition of the type of air contamination that must be counteracted.

Gases pass right through masks that are made to filter suspended particles from the air. By definition, gases are composed of molecules that are so small that they do not settle out by gravity. Gases may be changed to liquids or solids by reaction with appropriate chemicals, and some of them may be adsorbed by activated charcoal filters. Dangerous gases are often released when plastics and other materials are decomposed by heat, and bacterial action can also be responsible for their presence in excavations. Many gases are heavier than air and, if they are present in a place where there is little natural air movement and no forced ventilation, they may remain even when it would seem logical that they would diffuse into the air. Men have died in open pits only a few feet deep because they could not imagine there could be a dangerous concentration of a heavy gas in the depression.

Dusts are composed of solid particles that are large and heavy enough to settle out if the air is still. Grinding, sanding, and drilling dusts are readily recognized, and when overspray from painting evaporates, the solids in the paint become dust particles and the liquid becomes either a gas or a vapor.

Vapors are distinguished from gases by their ability to change to a liquid or solid when the pressure is increased or the temperature is reduced a moderate amount. Water vapor behaves like a gas until it encounters a cold surface and condenses on it as a liquid or frost. There is no need to make a careful technical distinction between gases and vapors at this point but, if working conditions are such that a vapor is apt to condense or solidify, it may interfere with the free movement of valves in a respirator or clog a filter at an unexpectedly rapid rate.

Fumes are dusts; the term implies that the dust is of a particularly objectionable nature and is frequently used to describe the air contaminants formed when very fine particles of metal are released into the air by the heat of a welding or burning operation on a plated metal. Fumes are often highly irritating or toxic and may be composed of such finely divided materials that they are difficult to filter out of the air.

Mists are composed of droplets of liquids and are often produced by spraying, splashing, or similar mechanical work that disperses a liquid into the air. Mists can also form when vapors condense in cold air just as fog forms in cold air pockets or where the air above a body of water is cooler that the vapor evaporating from the water. If the finely divided particles of a mist are caught by a respirator they may clog it rapidly and if warmed they may also evaporate and pass through the filter.

Bacteria and spores may be freely suspended in the air and be so light that they act more like gas molecules than dust particles. Many dusts and mists carry bacteria and when this is the case removal of the coarse particles will incidentally remove bacteria that would otherwise pass freely through the filter. A filter that is designed to remove bacteria may be so fine that it will be rapidly clogged by dust particles ordinarily considered harmless.

Proper selection of respirators can be difficult; it is very easy to overlook an important detail. Moreover, commercial products often contain several toxic substances whose combined effects may be impossible to predict. Direct experience with the mixture is essential in such cases. Therefore, even safety engineers who have a good background in toxicology will benefit from a discussion of the product with a specialist on the manufacturer's staff. That person may recommend a specific respirator or suggest several appropriate ones but in some instances he may be very reluctant to discuss the product and be willing only to send a contractor a material safety data sheet giving technical information. In such cases, you, as a prudent safety engineer, should call the manufacturer of the respirator you contemplate using and perhaps check with the local department of health as well.

I am reluctant to present the concept of the Threshold Limit Value

(TLV) because it is just one factor affecting the choice of a respirator. However, it is too important to omit entirely. The TLV is the concentration of a substance in the air that most industrial hygienists consider permissible in a work environment. Much more could be written here, including comments on the relationship between the TLV of a substance and the limitations that are appropriate for the same substance as an atmospheric pollutant to which the general population would be exposed 24 hours per day. Note that I am discussing protection of persons during their working hours, presumed to be about eight hours per day.

If the TLV is very low, it indicates the material is very toxic or highly irritating; a correspondingly effective respirator is needed. Many respirators are not good enough when the TLV is 0.05 milligram per cubic meter of air; they cannot be relied upon to protect the worker from injury by limiting the contaminant reaching his lungs to that low a level when the ambient air is more highly contaminated than the TLV.

But every respirator that is capable of meeting the TLV requirement is not necessarily suitable for every job. If the work area is very dusty, small diameter filters tend to clog rapidly, long before their ability to remove toxic substances has been exhausted. Similarly, respirators that use chemicals to trap vapors or gases lose their effectiveness when they have reacted with a certain quantity of contaminants. At that time the user must be able to detect the impurity and change the filter promptly. But some toxic materials are odorless and consequently they give the worker no warning. Reactive filters are dangerous to use as protection against such substances. There are also toxic materials that have very distinctive odors, but they also have the ability to dull a person's sense of smell. If a worker fails to change a filter as soon as he gets the first indication that such a contaminant is getting through his respirator, it will not be long berfore he can no longer detect the danger. For those types of contaminants, reactive filters should not be used.

Oxygen deficiency in the air is a very important hazard, and filters cannot protect workers from it. Self-contained breathing apparatus that includes a container of compressed air carried by the worker is required. Combination units are available that supplement the self-contained equipment with a hose from a suitable remote source of air.

Safety engineers should recognize that some people cannot wear respirators because, to be effective, they must fit closely against the person's face and leakage space around the shafts of spectacles, sideburns, and whiskers will make it impossible to get a satisfactory seal. Facial irregularities and missing teeth can also cause problems. An air supplied respirator will be effective in such cases provided the air supply is adequate in volume to meet respiratory needs and to maintain a positive pressure in the mask at all times to prevent any inleakage.

Eye protection must be considered along with respiratory protection because many substances that are toxic when inhaled are also irritating to the eyes. A worker wearing ordinary goggles may be unable to see. A full face type of respirator is needed in such cases.

NIOSH approvals for respirators designate the type of protection each will provide. Therefore the simple statement that a respirator has been approved does not necessarily indicate that it is suitable for any specific use; it must be approved for the intended application. This should be obvious, but it is surprising how many people think that any approved respirator can be used for any exposure.

Simple molded paper or cloth and paper or porous fiber face masks are inexpensive and are often sold as disposable single use items. Some are NIOSH approved, others are not. For nonhazardous materials such as coarse sawdust, just about any of them will make the worker's job easier, but only approved models should be used if toxic materials or asbestos fibers are in the air. For construction work where the mask may be bumped or punctured, I have little confidence in this type of respirator because it can provide a false sense of security. Some contractors supply them to employees because they are cheap regardless of whether or not they are suitable, and workers are seldom well enough informed to debate the boss's statements.

Most workers are familiar with molded rubber or flexible plastic face masks fitted with two replaceable round cartridges. Below and between the cartridge holders there is a valve that allows the wearer to exhale freely. When the cartridges become clogged with dust, inhalation becomes more difficult and this indicates the need to replace the dust filters. Cartridges are also made to trap mists, organic vapors, fumes, ammonia, acid gases, and some other specific contaminants. Often the cartridge holder will accommodate two types of inserts. The outer one usually removes dust, mists, and fumes while the inner one is specific for the type of toxic material in the air.

If the job produces a cloud of dust close to the worker's face and quickly clogs his respirator, consider using the type of breathing equipment that has a short flexible hose between the face mask and the filters, which can be clipped to his belt behind his back away from the source of dust.

Figure 3.2 shows a full face mask with two sets of cartridges. It protects the eyes from irritating materials and includes a rigidly supported adjustable eye glass frame as an optional feature. Prescription lenses may be mounted in the frame without requiring ear shafts that would interfere with the seal between the mask and the worker's face. The strap design does not interfere with the worker's hard hat suspension.

Figure 3.3 shows a full face mask with a hose through which air is supplied from a remote source or from the compressed air tank of a self-contained breathing apparatus. Prescription lenses can be mounted in it, if desired.

Respirators that rely upon the user's inhalation to draw air into the mask are typically permitted only if the TLV of the air contaminant is no lower

Figure 3.2 This full face respirator features dual cartridges and an exhalation valve located between them. The six-strap harness is designed to be worn under a hard hat. Note the rigid adjustable spectacle frame; it has no side shafts to interfere with a tight seal between the worker's face and the respirator. Appropriate cartridges are available for use where there is exposure to a wide variety of gases, dusts, fumes, mists, and organic vapors, but no cartridge type respirator may be used if the atmosphere contains less than 19.5% oxygen or is contaminated with dangerous substances that cannot be detected by the wearer when the cartridges have been exhausted. (Photo courtesy Willson Safety Products, Reading, Pennsylvania 19603.)

than 0.05 mg/m^3 or 2 million particles per cubic foot of air because there is always the possibility of slight inleakage of contaminated air when the worker inhales. To avoid this possibility, battery-powered air-purifying respirators are available that draw air in through high capacity filters and blow it into the face mask to maintain a positive pressure, thereby avoiding leakage around the mask when the worker inhales.

If the worker can escape from the contaminated work area without wearing a respirator, air can be supplied to an approved mask through a hose. If the hose is of suitable size and length, the worker simply inhales through it and exhales through a valve in the face mask. Worker fatigue is reduced, and longer hoses may be used if a blower is used to move air through the hose. Some blowers are actuated by compressed air, elimi-

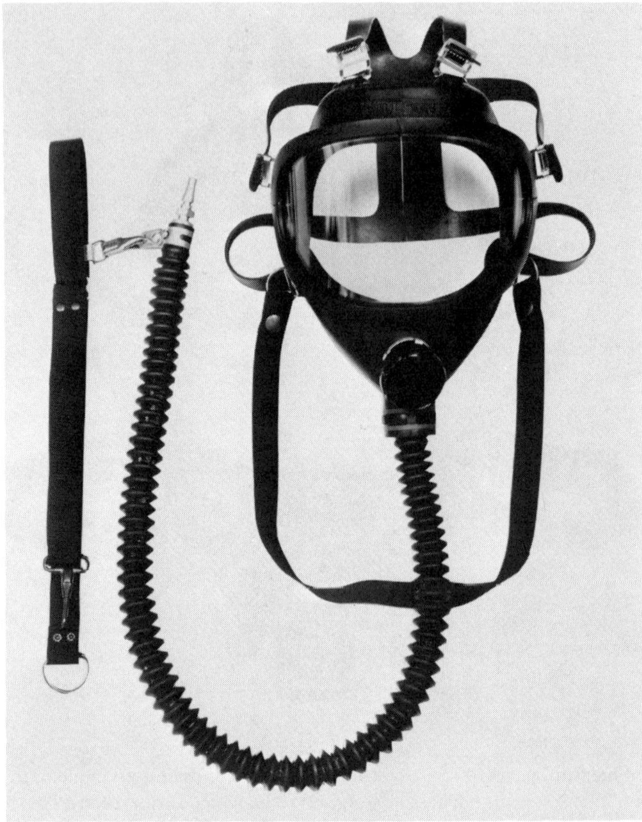

Figure 3.3 Air line full face respirators of this type are supplied with air through a hose in contrast to cartridge type equipment such as that shown in Figure 3.2. They are therefore suitable for use in many situations where the contaminant cannot be safely removed by cartridges or where the atmosphere is deficient in oxygen. Various means of supplying air are available, including portable compressed air tanks and either compressed air hoses or hoses that carry air at close to normal atmospheric pressure. Specific OSHA rules must be followed, as well as the manufacturer's instructions. (Photo courtesy Wilson Safety Products, Reading, Pennsylvania 19603.)

nating the need for electric power. Of course, the inlet to the hose must be placed where the air is free from objectionable impurities. If compressed air, free from any oil or other impurities is available, it can be supplied to the worker's breathing apparatus and reduced to atmospheric pressure by a regulator attached to his belt. This reduces the diameter of the air supply hose and eliminates the need for a separate blower.

When the atmosphere is deficient in oxygen or contains toxic impurities in dangerous concentrations, the worker's breathing apparatus must include a portable tank of compressed air.

OSHA regulations regarding respiratory protection in the construction industry are in 29CFR 1926.103, but OSHA has also identified a portion of the General Industry Regulations, 29CFR 1910.134 as applicable to construction and this latter regulation includes, by reference, the American National Standard Practices for Respiratory Protection, Z88.2-1969. All these rules may be of importance to safety engineers when workers need respiratory protection. They mandate that those workers who need respirators be properly trained in the use of necessary equipment. This is important; no respirator can be effective unless it is properly used.

Appendix G includes a table from 29CFR 1926 on the selection of respirators. Also, Section 7.7 on work in enclosed spaces includes comments on respirators.

3.5.4 Hand Protection

Cuts, blisters, and abrasions are common hand injuries. Also, when workers cannot get a good grip on objects, foot injuries can occur and other workers may be struck by falling material. The gloves available from local hardware stores are often more suitable for weekend gardeners than for construction workers. Firms that specialize in industrial gloves offer a wide variety of products, many of them designed for special types of work. Note how gloves are made; some have a heavy seam where the fingers are stitched to the palm and this is annoying to some workers. Other makers use a single piece of leather for the palm and the adjoining parts of the fingers. The quality of leather used varies greatly, as does the fabric and stitching. Gauntlets provide wrist protection, but they can also catch sparks and hot chips. It may be better to choose gloves with knitted wrist protection or with a strap, but such gloves cannot be shaken off quickly in an emergency. For some jobs, hand pads, mittens, or even finger guards are better than regular gloves. Resistance to heat or cold can be important considerations. For some jobs, gloves are needed to protect the work from corrosive moisture on worker's hands. Electrical resistance can be important.

When workers must handle chemicals hand protection can be a technically difficult problem. A glove that soaks up an irritating material and holds it in contact with the skin can be worse than no glove at all, and yet an apparently impervious glove that has a hole in it and a cloth lining can cause this problem. To resist wear and perforation, heavy duty chemical resistant gloves are available, but they may be very stiff, particularly in cold weather.

Molded natural rubber or latex gloves are pliable and relatively comfortable to wear. They have good resistance to tearing and abrasion, but they are a poor choice if there will be exposure to benzene, Freon®, gasoline, kerosene, diesel fuel, hydraulic oil, lacquer thinners, turpentine, and a number of other materials that will cause rapid deterioration. Neoprene synthetic rubber is better for exposure to gasoline, diesel fuel, oils, and thinners. It is better than butyl synthetic rubber for most of these materials, but may be inferior for handling certain chemicals. Buna N synthetic rubber is usually suitable for the same exposures as neoprene, but each type of synthetic rubber has certain qualities that make it the best choice for certain uses. Manufacturers publish lists of chemicals and the comparative resistance of different materials to them. Plastic gloves and plastic-coated cloth gloves are excellent for many purposes.

All the impervious gloves can leave a highly objectionable perspiration odor on worker's hands. Sometimes the best way to control this is to rinse gloves out at the end of every day and hang them up on pegs to drain and dry. An alternative is to have workers wear lightweight cotton gloves inside the impervious ones; in cold weather this can serve a dual purpose. A third alternative is to use heavy abrasion resistant gloves over thin disposable plastic ones.

Gloves coated with a high visibility orange plastic are excellent for signalmen. They not only make the hands easy to see, but they identify people who are authorized to direct traffic or to give hand signals to machine operators.

Glove selection is a good topic for tool box meetings, particularly if an assortment of types can be assembled and someone is present who is well qualified to explain the advantages and disadvantages of each type.

Hand protection does not stop with the selection of proper gloves. Vacuum cups are very useful for handling glass and sheet metal; they can speed up the work and avoid hand injuries at the same time. In the section on materials handling I have mentioned pipe sticks for lifting pipe from piles without the danger of crushed or cut fingers. Vibration white finger disease is another hazard (see Section 7.3). Tool selection is important in avoiding this hazard. Hand creams are made in two general types; one is for water soluble irritants and the other is for oily materials. When they are worked well into the hands and under the nails, industrial hand creams make it easier for workers to wash up and they can considerably reduce the tendency for workers to develop skin rashes.

3.5.5 Footwear

Many workers think the job is the place to wear out their old shoes, but if they fall because they were wearing loose fitting sneakers or if they have foot injuries because their shoes were paper thin, it is the employer who has to fill out all the forms, pay the medical bills, and do all the rest of the things that are required after an accident or occupational injury. Consequently, many employers try to encourage the use of safety shoes, even if they have to pay part or all of the cost.

Some manufacturers of footwear have salesmen who will come to the job in a van and fit workers there. They may even carry a stock of the most popular sizes and styles of safety shoes with them.

The American National Standards Institute (ANSI) standard Z 41.1

Figure 3.4 These three styles of cold weather safety boots feature warmth, comfort, and long wear. They have cushion insoles, anchor flange steel box toes, speed hooks, and are available with several types of outsoles and heels to meet specific job requirements. (Photo courtesy Hy-Test Safety Shoes, Division of International Shoe Company, St. Louis, Missouri 63166.)

recognizes three classifications of steel-reenforced safety shoes, designed to withstand impacts of 30, 50 or 75 foot-pounds without excessive deformation of the steel toe protection (Figure 3.4). Many styles are available; some are indistinguishable from dress shoes, while others are similar to typical work shoes. Workers often expect them to be heavy and uncomfortable but this is not the case if they are properly fitted. In fact, they are often more comfortable than regular shoes because they have more toe room.

Some safety shoes include flexible metal inserts in the soles to resist puncture by protruding nails. Other designs have high electrical resistance to protect workers who may accidentally touch live wires. Low resistance footwear is also made for those who work in explosive atmospheres and must not permit static electricity to build up on their bodies and cause sparking when they touch a grounded metal object. Wooden sandals are available for workers who must walk on hot surfaces such as freshly placed asphalt.

Safety shoes usually only shield the toes from falling objects, but when additional protection is desirable instep guards may be fastened over either safety shoes or conventional footwear.

Some contractors on the Alaskan pipeline recommended that workers prepare themselves for cold weather and mud by obtaining boots with a

Figure 3.5 For cold, wet, or muddy conditions, sturdy footwear such as this Maine cold weather boot is a good choice. It has a water-resistant steerhide upper, crepe rubber or synthetic sole and vamp, and a removable wool felt liner. (Photo courtesy L.L. Bean, Inc., Freeport, Maine 04033.)

rubber sole and vamp stitched to sturdy leather uppers and lined with removable felt inserts. This style is often called a Maine cold weather boot, and it has been popular with outdoorsmen for generations. Figure 3.5 shows a popular model with a crepe rubber sole molded to a ground-gripping chain pattern. The manufacturer will replace the rubber parts, stitching them to the original uppers, at a cost of less than half that of a new pair. Of course, these boots are not safety shoes, but they are good protection against frostbite and wet weather.

3.5.6 Hearing Protection

See Section 3.3.3.

3.6 Training Activities and Problems

Workers are most likely to be injured during their first month on the job. The Bureau of Labor Statistics reported that 2528 construction workers were injured in Maryland in 1976, and 23.89% of the cases involved people who had been on the job one month or less. During the succeeding 2 months, the monthly percent of injuries was only about one quarter as great.

Young workers in all industries are more apt to be injured shortly after they are employed than their elders. In the analysis of 5822 injuries in North Carolina and Iowa in 1977, all of which involved workers from 16 to 20 years of age, the Bureau found that 30.52% of the cases occurred during the first month of employment. For the 21 to 25 year old group, the same study reported first month injuries were 13.71% of the annual total. For the 26 to 30 year old group, the corresponding figure was only 8.47%, and for 46 to 65 year olds, it was under 4%. Clearly, it is important that safety training be started as soon as employees report for work, and that particular attention be paid to those having the least work experience.

First-day safety instruction provides an opportunity to double check the qualifications a worker appeared to have when hired. He may require special training if found less capable than he seemed to be when given the job.

Some job applicants are apt to be functionally illiterate. I have found it disturbing, after talking to a person about a job and finding him apparently qualified, to discover he cannot read or write. When asked to fill out an application blank, some people have simply gotten up and walked out. Sometimes, when I suspected a reading problem and asked the applicant to look at a newspaper advertisement, the applicant could not recognize either the name or price of any common commodity such as apples. Quite often these people have had an acceptably good command of spoken English, but had just never learned to read or write. I do not know the reasons, but perhaps some may have had an illness (dyslexia) that interferes with a person's ability to learn to read. Others may not have had an opportunity to attend school, but it is no credit to our educational system that some have attended classes for eight years or more without becoming sufficiently trained in reading and writing to be employable in other than the most unskilled jobs where they would be under constant supervision. Consider the danger of having a worker on a construction job if he could not read signs such as High Voltage; Flammable Liquid; Man Working Inside, Do Not Start; Emergency Exit; Beware of Guard Dog; Fire Extinguisher; Smoking Prohibited; Authorized Personnel Only; and Explosives, Keep Out.

I do not consider workers who can read and write a foreign language a comparable problem. Once a person has learned to write one language, he generally has no trouble recognizing danger and caution signs in English after being given a minimum of instruction. Bilingual signs and pictorial warnings can also be used where a high percentage of the available workers speak the same foreign language. They can also help train one another. The dangerous applicants are those who cannot read and who are adept at concealing that fact. If they have not learned to read by the time they are adults, I doubt whether on-the-job training can overcome the deficiency. Some communities have volunteers who help adults learn to read; I hope that they include warning signs in the vocabulary taught.

During a vacation trip to a foreign land, one quickly realizes the plight of the functionally illiterate worker. Safety engineers may encounter similar language barriers when they first attempt to understand the details of insurance policies. For example, in a typical crime policy, coverage for robbery restricts the insurer's liability to the value of the property, and any bodily injury must be insured separately.

I recently bought a travel disability policy to cover the costs incidental to returning home from abroad in case of accident or illness. Just before departure, I noted that the coverage was for a person in a "nonambulatory" condition only, which was defined to mean that the patient "is not able to walk, is confined to bed, and any transportation must be made in a fully horizontal position on a stretcher." "Ambulatory" was defined as any condition other than "non-ambulatory." In other words, no part of the traveler's homeward bound expense would be paid if he could be moved in a wheelchair but could not walk. I had always assumed that a person confined to a wheelchair was non-ambulatory.

Popular or common usage of terms is always superseded by the definitions in the policy. Moreover, the legal definition of a term may be different from either the common usage or the insurance policy definition.

A first encounter with the language of industrial hygiene can also be baffling until the technical meaning of common words is understood. I have attempted to explain technical terms in their context rather than in a glossary.

It has been my experience and observation of the experience of others that responsibility for safety may be assigned to an engineer before he has had any formal training in the subject. He is then expected to train himself and others, learning as he encounters problems. For those having similar experiences, the sources of information in Chapter 2, Section 2.3 will be useful. Attending a short course or seminar is one way to get a quick orientation in the safety field, but the field is such a broad one that it is difficult to be sure that the proper course has been chosen and that it will be worth the time and expense. The decision is a very personal one; some people find courses invaluable, others have been disappointed and prefer to go at a slower pace and dig out the answers as they are needed.

I suggest that newly assigned safety engineers discuss their problems with the insurance company that handles their worker's compensation. Such companies have a direct interest in reducing claims, and they are in a position to know about the types of claims that have been made by their client's workers. Insurance companies usually have money available for the promotion of safety, and they try to make the most cost-effective use of it. Return on money invested in helping a safety engineer should be attractive to them. If a new insurance policy is about to be purchased, consider the assistance various firms can and will provide in training foremen and workers. Assistance may be offered as an inducement to accept a policy.

Some firms have audiovisual training aids; they may have meetings that may be attended by safety engineers employed by their clients; others cooperate closely with program chairmen of technical societies and provide speakers. Most technical societies welcome visitors at meetings; ask your insurance agent about upcoming programs. He will probably know about any that touch upon the safety field, even if the speaker is a competitor. If audiovisual aids are available for use at on-site meetings, I suggest the short ones be preferred. They are easier to schedule and less liable to be crowded out of an agenda at the last minute.

Printed training aids are available from many sources. Check with manufacturers of equipment, trade associations, and labor unions. Most of us have to learn on the run, so the brief publications with "right" and "wrong" pictures are preferable for general educational purposes, whereas the technical bulletins, standards, and many government publications are preferable when a particular problem must be solved and every detail must be considered. In the descriptions of the sources of information in Chapter 2, I have tried to indicate the type of information that each agency is best equipped to provide.

4

Site Preparation

4.1 Site Inspection

In small firms, the safety engineer may also be expected to be aware of possible unexpected events that may not be potential causes of injuries but that are capable of causing substantial financial losses. His initial inspection and investigation of the site may reveal unanticipated problems.

A common occurrence is the discovery of chemicals that have been dumped on the site. It may have been done even after the contract for construction was signed; it can happen overnight.

Squatters often move into vacant buildings, and scavengers have been known to rip out plumbing to get metal they can sell for scrap. They may weaken the structure by removing metal columns or lintels.

Site inspection may disclose sink holes, springs, marshy areas, old wells, mine shafts, foundations, buried tanks, and even abandoned oil or gas lines. In wooded areas, an old burial ground may be found, dating from colonial days. The planning of a safety and loss prevention program must be based on knowledge of the site as it actually is, rather than on anyone's understanding of it that is based on statements by others. But even then, there may be surprises.

In Rocky Hill, Connecticut, when bulldozers removed the top soil, dinosaur tracks were discovered in the underlying rock. The site of a proposed building became a state park and the building had to be erected elsewhere.

In San Diego, California, close to the Point Loma Submarine Support Facility, excavation of a trench was halted when the remains of what appeared to be an eighteenth century Spanish fort were disclosed. This involved federal regulations regarding archaelogical investigations.

In New York City, construction of a 30 story office building was interrupted when excavation workers ran into the remains of an 80 foot long merchant ship that had been buried since about 1750. Archaeologists agreed to reclaim only part of the ship, and the owner of the building agreed to hold up work for about a month. He estimated the delay would cost about half a million dollars.

Local historical societies often become concerned when they learn that an old building is about to be demolished, and unfortunately they are apt to become active only a short time before work is scheduled to begin. Their frantic efforts to save the structure often delay work while attempts are made to raise money to move the building or buy it and the land.

It is becoming common for local zoning regulations to require an archaeological study of every site by a person with a college degree in that subject every time a piece of property is to be subdivided or a new building is to be erected.

Everybody has read about large jobs being delayed by environmentalists who are concerned about the effects of a dam or power plant on the ecology of the area. Somewhat less publicity has been given to laws governing the filling of wetlands, which may include areas that were once tidewater zones but have not been submerged at high tide for years. Wetlands may also include marshy ground or flood plains far from the seacoast. I suggest that local laws be investigated before work begins in order to avoid the possibility that an injunction may stop work and tie up heavy equipment.

The way these initial problems are handled can affect community relations throughout the time the building is under construction and even thereafter. Riding roughshod over the sensibilities of local people, and particularly influential oldtimers, may not be good business even though the immediate cost studies show otherwise. People who hold positions on the boards of local historical societies are often wealthy and influential officers of corporations, or are in a position to control other projects in ways that are impossible to trace. Some little old ladies can be harder to beat than a tiger in the dark.

4.2 Protection of the Public: Restricting Access to the Site

As soon as a contractor signs a contract, he may be named as a party to a suit if anyone is injured at the site. Therefore it is prudent to fence it in and to keep unauthorized people off the property.

Why do people want to get onto a construction site? Among many reasons, the rise in value of firewood makes any tree that is not protected attractive to freelance loggers. Old buildings attract history buffs; some look for old wall paper or paneling, carved fireplaces, hidden hardware, old bottles, old power line insulators, or antique car parts, to name just a few items. Some collectors, including dealers, have no respect for property or safety. They are simply out to get whatever they think is salable regardless of what damage they do to other parts of the building.

When excavation begins, you can expect visits by mineral collectors. Some are likely to be very inexperienced and totally unaware of the dangers involved. Others may be professional geologists or minerologists, seriously interested in the exposed formations.

Some contractors are totally unreasonable in their refusal to admit anyone to a construction site, and there is often a good reason why a person wants to make an investigation. It is my recommendation that requests be required to be made in writing, stating the purpose of the proposed visit, and the qualifications of the writer. If it seems that some legitimate purpose may be served by granting the request, set a time at your convenience, and get a signed release from the visitor that makes it clear that you are absolved from responsibility for his safety while at the job. You may want to request that the visitor be accompanied by one of your employees and, if so, it is appropriate that the visitor pay the wages of his guide.

Simply fencing in the property and posting "Keep Out" signs may be sufficient to keep out intruders, but in many places it is advisable to have guards, guard dogs, or an alarm system, particularly when a building is being renovated. Some alarm systems use light beams; interrupting any beam sets off a siren. Other systems that are moderate in cost and easily moved from job to job use miniature radio equipment. Disturbance of any one of a number of remote sensors triggers response at a master control station. In Scotland many of the distilleries have been surrounded by two fences, with flocks of geese free to move about between the enclosures. Local residents claim that they are far more effective than dogs; anyone trying to invade their territory is attacked and the squawking can be heard for a long distance.

The time to set up a safety sign is as soon as a permanent spot for it can be selected. The sooner workers can be made aware of the safety program, the better. OSHA requires that a poster, their form 2203, be posted at every site. The sign might also include a copy of the firm's safety policy statement. The sign used by Bechtel at Philadelphia Electric's Limerick Generating Station was shown in Figure 2.1.

4.3 Demolition Work

During demolition of a three story building in Sanford, Florida, 10 employees of the same contractor were looking at a faulty table saw inside the roofless shell when a gust of wind toppled a 34 foot high brick wall. Three men were killed and three more were seriously injured. The incident is a typical illustration of the importance of maintaining structural stability at all times during demolition work.

One Sunday morning in June, 1982 a man was trespassing on the roof of a one story building that was being demolished in Manhattan. The structure collapsed, tossing him into a pile of rubble and injuring his back. He was one of the luckier ones; many similar accidents occur every year and fatalities are not uncommon.

Everyone has probably noted similar news stories about old buildings that have collapsed without apparent reason. Often removal of adjacent structures is a contributing cause, and sometimes remodeling work done by incompetent contractors has weakened the structure. Fire, water damage, corrosion, or rot may have weakened critical parts of the building, leaving it in precarious condition. Demolition workers may therefore be walking into unsuspected hazards. Employers should recognize that demolition work may be the most hazardous part of a project. Certainly, it is the most unpredictable stage.

During demolition, it is obviously necessary to protect the public from falling objects, but it should also be remembered that when a building is being torn down it is particularly attractive to many people. Scavengers come looking for scrap metal they can sell, home owners see a chance to get bricks or wood for their do-it-yourself projects, and some people look for conversation pieces such as tiles, decorative stone carvings, and iron work. Treasure hunters may hope to find something of value that has been hidden away under floorboards and forgotten. Under "attractive nuisance" laws, contractors may be liable for damages if such uninvited visitors are injured.

OSHA regulations in 29CFR, Subpart T, paragraphs 1926.850 to 1926.860 inclusive should be read carefully by anyone in charge of safety during demolition work. A few of the highlights are:

1. An engineering survey must precede demolition, and the employer must have written evidence that it was made.
2. All service and utility lines must be shut off or capped outside the building line and any utility companies affected must be notified before demolition begins.
3. During demolition, the structure must be capable of supporting workers and debris safely. Specific rules deal with shoring and reenforcing work areas.
4. Each story, except for structural steel framing, must be completely demolished and dropped into the material storage space before work on the next lower story is begun. If mechanical demolition methods are used, rules for balling and clamming or the use of explosives must be observed.
5. Structural steel must be dismantled tier by tier in sections not exceeding two story lengths.
6. Masonry walls must be left in stable condition at the end of each shift.
7. Floor openings and chutes must be properly guarded.
8. Hazardous materials must be removed and any tanks or piping that contained them must be purged before demolition begins.

Old buildings usually contain pipes or boilers insulated with asbestos. Be sure to read 29CFR 1910.19(a) and 1910.1001 regarding exposure to asbestos dust. These references are general industry requirements that have been identified by OSHA as applicable to construction, and they are too detailed to quote in their entirety, but they are important. Use water to avoid creating dust if workers are handling asbestos in any form.

From 1976 to late in 1983, regulations limited the permissible concentration of asbestos fibers in the air to two per cubic centimeter and required that every fiber longer than 5 micrometers (0.0002 inches) be counted. In November 1983 OSHA issued an Emergency Temporary Standard, reducing this limit by 75%.

When dust cannot be controlled within permissible limits type C supplied air respirators must be used. Simple cartridge masks are not acceptable protection. When workers are exposed to asbestos dust, employers must supply protective clothing, and require that it be used. Contaminated clothing and all waste capable of releasing dust must be handled in sealed containers with prescribed warning labels. Work areas must be restricted to authorized employees, and caution signs must be posted at all entry points. Certain medical examinations and air quality tests are required. Records indicating the exposure of workers to asbestos dust and medical records must be available to workers; in addition, current regulations require that employers retain records for 20 years. These rules are being questioned by many contractors at the time of my writing, and I suggest that the latest OSHA regulations be checked before work is begun; changes may have been made by that time.

Old buildings may have been used for many purposes since they were erected, and the residues in ducts, under floors, and in pipes may be dangerous. Some accidents have been caused by explosion of compounds formed by the reaction of sodium azide with metals. That chemical is used in hospitals and clinical test laboratories, and it is not only a poisonous substance by itself, but the compounds that form when it reacts with metal in pipelines may remain there until detonated by a sudden shock. The same chemical may be used in automobile air bags and scrap metal processors have been concerned about possible hazards when autos are junked with air bags in them.

Old buildings may contain partly filled drums without labels; they may contain gasoline or other flammable liquids. A white granular substance may be any one of many substances, including caustic soda, a strong alkali that causes severe burns. I recommend that contractors call firms that specialize in the disposal of chemical wastes and let them remove all such chemical containers. The local department of health or the Environmental Protection Agency (EPA) will undoubtedly know how to get in touch with an approved firm. If the building ever housed a medical lab, the disposal firm will probably know how to decontaminate the drains before demolition starts.

If demolition workers are first on the job, that is the time to insist that everyone wear a hard hat, including supervisors, visitors, and inspectors. Keep a few distinctively marked hats on hand for people who are not regular workers at the site. Sometimes a color code system is advisable; it can be used to identify supervisors, fire watchers, or any group whose members might need to be spotted at a distance in an emergency.

If it is necessary to send anyone into a basement, pit, tank, or other restricted space, remember that the air may be unfit to breathe. Gas may have pocketed there, or there may just be insufficient oxygen in the air. Make the area gas free first. (see Section 7.7).

Demolition workers should wear safety shoes with steel toes and thick soles. Their work is more hazardous than that of most other trades with respect to the possibilities of foot injuries; special arrangements may have to be made to be sure they have the proper footwear. Sometimes the contractor pays all or part of the cost of safety shoes.

4.4 *Use of Explosives*

OSHA regulations governing blasting are in 29 CFR Subpart U, paragraphs 1926.900 to 1926.914, and also in paragraphs 1910.109. These regulations also refer to some industry standards, such as the American Table of Distances or Quantity Distance Tables dealing with the permissible quantity of an explosive that may be used, based on work done by the Institute of the Makers of Explosives. Rules of the Internal Revenue Service regarding storage of explosives are contained in 26CFR part 181. Transportation of explosives is governed by a number of Department of Transportation rules, the most commonly applicable of which are in 49CFR parts 390 to 397, dealing with motor carriers. State and local regulations also apply, and they may be more restrictive than federal regulations. For example, movement of explosives through vehicular tunnels or over certain bridges may be prohibited.

Under 49CFR, paragraph 174.20(b) the Bureau of Explosives of the Association of American Railroads has certain responsibilities for recording and publishing regulations for the transportation of hazardous materials. The Bureau publishes and keeps current a detailed compilation of DOT regulations applicable to all modes of transportation, not limiting itself to rail shipments. This book, Tariff BOE No. 6000, is available in loose-leaf form, and revisions are issued frequently. Anyone concerned with the movement of explosives would find it useful.

Every one of these regulations may be of critical importance at your job, and it would be unwise for me to attempt to summarize them; a vital part might easily be omitted. For this reason, explosives should be used under the direction of a professional blaster. Be sure to choose one who has experience in the type of work that has to be done at each job. A specialist in rock removal is not necessarily qualified to demolish a masonry structure. Check your blaster's experience as well as his license.

Before explosives are bought and delivered to the job, the blaster will want to participate in the purchasing, transportation, storage, and security arrangements. He will be knowledgeable about traffic control, warning signals, and preparation of drill holes, and must approve of the handling of such matters. After every blast, he must check for misfires and give his approval before others reenter the blast area.

Of course, the blaster is in charge of safety at the time of a blast, since he must be certain that everyone is clear of the area before he sets off the charge, but he cannot be everywhere at once, and it is important that all concerned be aware of the coordination plan that has been worked out.

In addition to the technical and legal preparations for a blast, some public relations and loss control work should be done. Tell local residents what is going to happen. Let them know when noise or vibrations are to be expected. Newspaper announcements may be appropriate as well as use of the local radio station. When planning blasting near existing buildings, discuss the work with your insurance advisor. He may want to go with your representative to inspect all property that might be affected.

Inspections have several purposes. They may be needed to be sure no old or deteriorated buildings are apt to be damaged. They provide an opportunity to allay the fears of residents. They are your opportunity to anticipate and prevent unwarranted claims. Be sure to note cracks in windows, masonry walls, and plaster, and go on record with property owners that such conditions were present before the blast. Get the advice of your insurance company regarding the need for inspections following the blasting to confirm that no damage has been done or to settle any claims that may be justified.

Some contractors leave a brief description of the project with the occupant or owner of a building at the time of the first inspection. It should describe the warning signals that will precede and follow blasting.

Timing of blasts can be important. During mid-morning hours when people are active, a noise is much less apt to be disturbing that when they are at dinner or when small children have been put to bed for naps or for the night. Also, try to avoid blasting when it will be necessary to block traffic during morning or evening rush hours. The initial survey will probably give you an idea of when the work will be least disturbing; try to adapt your work schedule accordingly.

If a blast will be spectacular, consider arranging for people to witness it from a safe vantage point and help them get set to take pictures at the opportune moment. What could be a public nuisance can be turned into a festive occasion if it is handled with some imagination.

4.5 Clearing Work

If there is much wooded area to be cleared, the workers are probably employees of a subcontractor who specializes in that type of work and has trained the crew. Moreover, he probably has the right equipment and keeps it in good condition in order to maintain a fast pace. The amount of brush produced makes it mandatory that some plan for disposing of it be part of the operation. I am less concerned with safety in this situation than when a few trees have to be taken down and a general contractor chooses to do it with his own force.

Chain saws have become so popular in recent years that most workers have used them at one time or another to cut up some firewood or to take down a small tree. They think that their limited experience qualifies them as experts, but they are not necessarily prepared to handle a large saw and to take down big trees. Felling a tree where there is insufficient room for it to fall to the ground is dangerous. If it hangs up against other trees, it can roll as branches are trimmed off, trapping the amateur woodsman. Also, experts can predict where a tree will fall, and how the cut end will move off the stump. They know when to step aside, and leave themselves room to do it. It all looks easier than it is in practice.

Trimming off branches and cutting up the trunk with a chain saw are dangerous operations because the stress on the wood being cut causes it to bend as the cut nears completion. The blade of the saw may be caught, causing the chain to break and swing around the workman. Kickbacks are also a danger, particularly when the tip end of the saw is used to trim off small branches. Old trees often contain embedded metal that can break the chain. Fence wire, nails, and bullets are apt to be deep in the wood and completely hidden by new growth.

Chain saws have also replaced axes and brush hooks so extensively that it is hard to find anyone skilled with an axe. Enthusiastic young men are apt to flail away at a small log, wearing themselves out and accomplishing very little. As they get weary, their aim gets poor and they are apt to miss the log completely, snapping or cracking the handle. When the axe head flies off the broken handle, anything can happen. Giving an inexperienced worker a machete to clear brush is like giving a child a loaded revolver. If he is lucky, he will get blisters and use a hand saw or lopping shears before he lops off part of his body with the machete.

Getting rid of the slashings is hard work unless a chipper is available and even then there is a lot of carrying and cutting to be done. Feeding brush into the chipper hopper requires care; it is easily possible to be drawn into the blades. However, the danger is so obvious that there are relatively few recorded instances of fatalities. Letting brush pile up until it is tinder dry is of course a major fire hazard and must not be tolerated.

Poison ivy and other poisonous plants are unknown to a surprisingly large part of the work force. Even in Connecticut where it is about as common as crab grass I estimate that at least a third of the younger workers cannot recognize it. Some old timers claim to be immune to it, and as a result they encourage new employees to ignore it or even chew a few leaves.

The result can be serious poisoning and a disabling rash. Failure to get a reaction after initial exposure does not prove immunity; people can get a bad rash who never had any trouble before.

Most people are afraid of snakes, yet in a study of 460 fatalities two and a half times as many deaths were caused by stings by hornets, bees, wasps and spiders as by snakebites. Once a person has been stung by one of these insects, he may become allergic to the venom of the rest. Therefore, it is advisable to get an immunization shot from a doctor after being stung the first time. I recommend that wasp and hornet spray be kept on construction sites until the clearing work has been completed; some varieties of bees live in underground burrows, and I have learned the hard way that they can be very aggressive. Insect repellants are sometimes useful; I have found them only moderately effective, but worth trying.

In warm weather, ticks and fleas may be annoying and they can also be carriers of disease. A close friend's son died of Rocky Mountain spotted fever, following a tick bite. If workers wear infested clothing home, the vermin may multiply so fast that it may be necessary to fumigate the premises. Don't let them get out of hand; a serious infestation is difficult to correct.

4.6 Trash Disposal

On jobs where demolition is necessary, waste disposal is an immediate necessity; even a few days delay makes it difficult to avoid accidents when workers step on protruding nails, trip over loose material, or flammable material piles up. The time to establish good housekeeping procedures is before refuse piles up. Crack down on offenders immediately and insist that waste be carted away, put into containers, or handled in whatever other way may be agreed upon. If any contract is not clear, and there are disagreements about removal of rubbish, the longer it is allowed to collect the harder it will be to resolve the conflicts. If several subcontractors all dump into one pile, they are certain to argue about how much each should contribute to the cost of the cleanup job. In the end, the general contractor may have to pick up the tab. I suggest that a regular cleanup schedule be negotiated, so that at the end of every shift or the end of any other appropriate interval of time, the whole area will be cleared of debris.

If any explosives are found to be unfit for use, the blaster should supervise their disposal. He should also control the disposition of all packaging materials that were in contact with explosives, and supervise the cleaning of contaminated floors in magazines.

Empty containers that held any hazardous material must be handled as though they were still full until they have been thoroughly cleaned and purged of residual material. Watch particularly for drums that contained any flammable liquid. Traces of the liquid will form a mixture of vapor and air that will be explosive if it contains more than enough vapor to exceed the lower explosive limit, and less vapor than required to exceed the upper explosive limit. Since there is no practical way of predicting what vapor-air mixture may be in an empty drum, it must be assumed that the mixture is explosive until the drum has been cleaned. For some chemicals, such as alcohol, the explosive range is quite broad; any air-alcohol mixture between 7 and 36% alcohol can be detonated by contact with a spark or flame. For regulations governing transportation of empty drums that contained hazardous materials, see 49CFR173.29 which covers the rules established by the Department of Transportation. Trucking firms can be expected to have a copy of this regulation. If a number of drums are to be carted away and they are in good condition, it may be possible to get a company that reconditions used drums to come and get them.

Disposal of residual amounts of hazardous materials can also be a problem. If the product kills fish or wildlife in a pond or stream near the site, the Environmental Protection Agency will be investigating the source. If the material interferes with the sewage treatment plant's operation, someone will also be around to see who is responsible.

When any large system has been chemically cleaned, the detergent solution may create a disposal problem. Before beginning the operation, consider what will happen when the system is drained and flushed. I recommend that work of this nature be subcontracted to a company that specializes in it. Specifications may call for the use of solutions that are unacceptable to authorities have jurisdiction over pollution; it is the business of cleaning specialists to know what to do in such cases. Similar problems can be encountered when it becomes necessary to drain and flush a large cooling tower and the condenser cooling water has been chemically treated. Disposal problems often control the selection of the treatment chemical and specification writers do not always know what is or is not acceptable to local authorities. It pays to check before adding anything to a large volume of water such as a chilled water, condenser cooling water, or boiler water system. Watch particularly for chemicals that are yellow or orange colored; they probably contain chromates that are the target of many environmentalists.

Loose material that can be blown about by a high wind must not be allowed to become a hazard to workers or the public. Many injuries have occurred when glass, sheet metal, and empty containers have been picked up by the wind. Get such items into a trash container promptly , and if the weather looks threatening, make a special effort to police the job.

Dust has to be considered a waste product. Asbestos dust must not be allowed to become a health hazard to workers or to the public. See Section 4.3 of this chapter. Do not allow anyone to breathe the dust from broken fluoresent light tubes, and remember that some dusts can cause silicosis. Industrial equipment suppliers have dust collectors that fit over open-head drums. They should be more frequently used because they not only save cleanup time, but reduce the need for workers to wear respirators or dust masks that slow them down, assuming they continue to wear them when the boss is not watching.

When a contractor is hired to take trash to a dump or incinerator, it pays to keep an eye on him occasionally to be sure the trash is actually going where it is supposed to go. Most municipalities charge such contractors a fee for every load they dump, and there is always a temptation for them to make a detour to a vacant lot or ravine. If so, the contractor whose trash is found where it should not be will be held responsible and the chances of recovering the cost of the cleanup job are slim. Trash attracts trash, so that when one firm's refuse is identified, there will probably be a lot more in the pile by the time the local authorities require it be removed.

The same problem can exist in reverse. If a construction site is not fenced and is allowed to be visibly cluttered with refuse, others are likely to add to the piles. Sometimes the source of such foreign material can be traced by checking for papers or lables, but at best it is a time-consuming job.

4.7 Underground Obstructions

Before excavation work is done, determine whether any underground installations are likely to be encountered. There could be buried oil, gasoline, or septic tanks at some sites. If gas, water, electric, or telephone service lines are known to be present, OSHA requires that the utility companies be notified before excavation work beings. Approach the anticipated location of every buried obstruction cautiously. It may not be even close to the spot shown on the plans.

When excavating near an existing building, consider the possibility that it may be unstable, so that even a small amount of vibration can cause it to collapse. In the past year I have noted that several buildings in New York City have simply collapsed for no apparent cause, and without any construction work being done in the vicinity. Any contractor who might have been working nearby would certainly have been blamed for the damage, whether it was his fault or not.

The danger to an existing structure is of course particularly great if a cut must be made deeper than its foundation. Such work must only be done under the control of professional engineer who is experienced in such problems.

4.8 Surveying and Staking Out

While surveyors and engineers are staking out the site, it is necessary for them to concentrate on their work, but they sometimes forget that operators of earth-moving and excavating equipment are also intent on what they themselves are doing and may expect others to get out of their way. Once a surveying instrument has been set up and adjusted, it is not reasonable to expect it to be moved every time a machine operator wants to pass near it. This mutual disregard of the near presence of the other results in collisions that are harder on the surveyors than on the machine operators. Post flagmen if surveying teams are likely to be in danger.

The use of lasers for running lines involves hazards to other workers because the concentrated light beams they produce can be injurious to the eyes. Laser operators must be fully qualified, and must carry proof of their qualifications whenever they are using the equipment. OSHA regulations 29CFR 1926.54 include technical restrictions on permissible light intensities and safety precautions that must be observed.

Unless the safety engineer intends to become a qualified laser operator himself, it is advisable to make the person in charge of the equipment responsible for implementing the safety precautions. In order to do this, it is necessary to make all other workers on the site aware of the fact that authority has been delegated and that instructions issued by the laser operator have the same importance as those of the safety engineer himself.

Laser operations make a good topic for safety meetings. They are sufficiently new to be interesting, and new applications for them are constantly being developed, such as the automatic operation of graders in producing perfectly level fields of great extent where irrigation by gravity flow of water is necessary to grow crops. The whole meeting need not be devoted to lasers, but the topic does merit coverage if the equipment is used at the job and it is sufficiently new to liven up a meeting at which the only other topics to be discussed are ones about which everybody has heard before.

5

Excavations, Utilities, and Foundations

After a site has been prepared and before the above-ground structure begins to take shape, some of the operations most likely to cause accidents have to be completed. Among these are excavation work, rough grading, and possibly pile driving. At this stage of the project, accidents associated with the use of heavy equipment are likely to occur. At the same time, the safety engineer should anticipate that there will be a substantial change in the nature of his own job. There will be more workers and more subcontractors on the typical job and he will have correspondingly more activities to coordinate. His initial role as a planner will have to shift into a more active stage. I suggest that if you are in charge of the safety program, that you deliberately increase the time during which workers see you checking safety hazards. You will probably want to wear a distinctively colored hard hat. I suggest that if air quality is being tested in an enclosed space, that this fact be apparent to the workers who are affected; they may be interested in seeing the testing equipment. Anything that spurs their interest in the safety program is desirable. Keep them thinking about safety. But keep the first-line supervisor's responsibilities in mind; he may not appreciate your taking his men away from their work to look at your equipment. If he wants the tests made before his crew gets there in the morning, by all means get in and get out so he can get his job done. He will be more cooperative with you if you respect his wishes.

To increase your visibility you may have to change your customary work schedule to be where the action is at the moment. As a young trainee, I recall having been directed to be on hand at 4 o'clock on a Sunday morning to observe the start up of a steel mill operation. When I arrived, somewhat bleary-eyed and perhaps a little resentful, the work was proceeding on schedule and both the vice president in charge of operations and the chief metallurgist had been there for some time. The workmen were well trained, but it was possible for many things to go wrong and I think they appreciated the fact that the responsible executives had taken the trouble to be there just in case a quick decision had to be made. In the ensuing 10 years, the company had many troubles, but I never heard a derogotary remark about either of those men in all those years.

Personal observation of important stages of the job is important if there is likely to be disagreement about the safe way to perform the work. Differences of opinion are normal, but disputes about what was actually done should be avoided. I have never had a serious problem finding the safe way to do a job when the foreman and I watched it together; we could always work something out. But I have had trouble when I got my "facts" secondhand. The fat was really in the fire if the person to whom I was talking got the idea I was calling him a liar when I was simply stating what I honestly believed to be a simple matter of fact.

These comments about the relationships between people are offered because it is during the excavation and foundation preparation stages of construction that the first challenge to the authority of the safety engineer is likely to occur. If you have responsibility for safety, presumably your duties and your authority to carry them out were explained at a meeting or in one of the contract documents, but the first time a safety rule costs extra money or a customary way of handling a part of the job is challenged, what was said at a meeting can be very quickly forgotten. Handling the first conflict is critical; you had better be sure you are right before throwing your weight around. Once you make your move, you are putting your future status on the line.

This is the time when diplomacy and a firm hand are both needed. Backing a foreman into a corner, particularly when his men are listening, is sure to start a battle. Even when he is wrong, knows he is wrong, and knows your way is workable, the foreman cannot afford to lose face in front of his men. Nobody wants an accident, and even when I have had serious disagreements, it was usually possible to work something out that was mutually acceptable. Sometimes it is possible to compromise by delaying enforcement of a rule until the need for it is more apparent. It is hard to sell the idea that everyone should wear a hard hat when there is nothing but blue sky overhead, but when the first crane is scheduled to appear, that may be the day to announce that thereafter, hard hats will be required at all times. Everyone may not agree with the rule, but they can make enough sense out of it to go along with it. If there are several subcontractors on the job, have them wear color-coded hats. Ease of identification may help gain acceptance for the rule requiring that hard hats be worn.

Of course, timing of discussions is important; problems can often be resolved over lunch and a couple of beers that would seem much more formidable when one party has other things on his mind and is under pressure. If both people report to the same boss, it is usually best to keep him out of the discussion as long as possible.

If you, as safety engineer, have exhausted all other approaches and see no alternative to issuing a direct order that will be resented and possibly disobeyed, be sure that everyone up the line who might countermand your order understands the issue and will back you up. Sometimes your boss may disagree with you and know a better way to handle the problem. Bosses sometimes do know such things; that is why they are bosses. If, after thinking it all over, you find that it will be impossible to sustain your position, it is better to change it and avoid ending up with egg on your face. Remember that it is your job to avoid accidents. Whether you do it with tact, diplomacy, persuasion, or a sledge hammer is up to you. Just try to avoid blundering into a hornet's nest without recognizing it is there.

5.1 On-site Traffic

Whenever machines are moving about, there are many opportunities for accidents. In a British study of 100 consecutive fatal accidents at construction jobs, 15 involved excavating equipment or trucks. In 7 of the 15 accidents, a machine tipped over, fell into an excavation, slid into water, or got out of control in some manner. In every one of these accidents, the operator was at fault, although mechanical trouble was a contributing cause in two of the seven cases. Four more of the 15 accidents involved fatalities because the machine operators did not see another worker. In the remaining four accidents, one unauthorized rider fell off a truck; an excavator knocked over a pole and it struck a bystander; an operator got off his machine, leaving the motor running and it ran over him; in the last incident, a truck driver was killed in a collision with a car.

Injuries are frequently caused by material falling from a bucket or truck. Workers should be kept aware of the importance of staying out from under any moving load, or following too closely behind a vehicle from which anything could fall. Before any truck leaves the site to travel on the public highways, the load should be checked, but hardly any motorist will have trouble recalling having seen a truck from which dirt or rocks were falling onto the road.

Every operator knows that a front loader can tip over if it is braked too suddenly with the load raised, or if it hits a snag at excessive speed, but judgment is needed to select an appropriate speed and to keep the load high enough to clear obstructions without unduly reducing stability. No safety engineer should substitute his own judgment for that of the operators, but one good man can set the pace for the others. Developing judgment on the part of workers is difficult but important. Sometimes someone from the equipment manufacturer's plant can be induced to demonstrate safe operating practices at the job; his example may be beneficial in curbing the cowboys without slowing down the job below a reasonable speed.

If it becomes necessary to tow a machine that has become mired down, the work should be supervised by someone who recognizes the danger to anyone within reach of the end of a chain or cable if it snaps under a sudden strain, and the possibility that either the mired machine or the one attempting to tow it may move in an unexpected manner. Workers who gather around to watch are as likely to be injured as those actually involved in the job.

Someone has to establish a pattern for movement of materials and equipment on the site; if it is not planned, it will be done by every driver and machine operator to suit his own convenience. The result can be chaotic, and those who were able to commandeer the preferred storage areas may be very difficult to dislodge if there is no initial plan. This is a problem involving operating efficiency as well as safety; it makes little difference who solves it, provided someone does it who recognizes the importance of leaving room for on-site traffic to move safely.

5.2 Earth-moving Equipment

When a machine arrives at a site on a trailer, unloading is a critical operation, but it has to be done at a time when the least may be known about peculiarities of the machine's response to controls. Moreover, after being out of service since the previous job and after being bounced along over the highway on a trailer, even a machine that has always been easy to handle may not be as responsive as it was when loaded onto the vehicle. While unloading, take it easy; don't rush the job even if it becomes necessary to pay overtime.

If the machine is new, be sure to get the instruction manual and don't be too proud to read it even if you think you know all about it. Whenever a design change is made, the manufacturer may insert a new instruction or cautionary note. If, after operating the machine and having trouble with it you complain to its maker, his first response may be to ask whether you have read a certain part of the manual. Most manufacturers stand behind their products, but once in a long time a design defect is concealed by fine print in the manual. I once ordered a machine on which the electrical controls were inadequately protected from water and, when it was washed with a hose, several thousand dollars worth of damage was done. Failure to read the hidden instructions cost us most of the usual repair charges, and it cost the manufacturer the opportunity to sell any additional equipment to the company. Such incidents are, of course, the exception, but one experience is sufficient to make a person look carefully for any indication that a machine may have unusual characteristics. Before loading a machine onto a trailer, note whether the manufacturer recommends backing it up the ramp or driving forward onto the trailer. It may make no difference, or there may be a preference for one or the other method.

Before unloading a machine, check it over; if it is not in usable condition or if parts are missing, the time to object is before unloading. Don't sign for missing safety items such as a fire extinguisher, rollover protection, backup alarm, and special tools that have been ordered for doing hazardous maintenance work. Get all sets of keys.

When ready to unload, be certain the trailer and ramps are resting on a solid surface and are securely blocked. Check for blocks or anything else on the trailer that could deflect tires or crawler tracks. Watch for ice or mud. Have an experienced operator at the controls and remind him that the machine may not be in perfect condition. Keep everyone else out of the way so that if anything goes wrong, bystanders will not be in danger.

As soon as possible after unloading, put the machine through its paces. Make a thorough test; if it has to go back to a repair shop, the sooner a defect is discovered the sooner it will be back on the job.

Encourage every operator to have a sense of personal responsibility for whatever machine he is using that day. Be sure he has full information about its capacity and peculiarities, such as a tendency to stall, skid on slopes, or bounce at certain speeds. Ask him to report any additional features about which future operators might want to be told. Allow time for operators to check their equipment before starting work, and to make a trial run to be certain all is well. Pay attention if an operator feels that something is not quite right, even if he cannot say just what is wrong. A person at the controls may hear or feel vibrations that others cannot detect and if an operator's "sixth sense" tells him something is amiss, he is probably right. Of course every operator should walk around his machine before starting up; someone may be trying to retrieve a tool that has rolled under the machine, or be catching a nap in its shadow.

Don't allow anyone to climb on or off a moving machine or to leave one with the engine running and nobody at the controls. Never check anything without someone at the controls if the engine is running.

Never leave a machine unattended with the key in the ignition; an intruder or unauthorized employee may try to operate or move it. At the end of every shift, unless there is another operator ready to take over immediately, place and lock equipment so that if anyone touches its controls it cannot roll down a grade, and so that no boom, bucket, or blade will move by gravity if a control is moved. If necessary, use extra locking methods such as a chain and padlock. Don't rely on brakes; chock tires or tracks or place the machine so that it has no tendency to move even with the brakes off.

OSHA has specific instructions regarding the installation and testing of rollover protective devices. See 29CFR 1926.1001 and 1926.1002. See also 29CFR 1926.602(a)(2)(ii) and (iii) regarding seat belts.

Rollover accidents are usually caused by operator error or lack of foresight. If a machine seems about to get out of control, applying more power or turning sharply may be the right or the wrong thing to do at the moment, and the "cause" of an accident is often said to be an operator error of this nature. In-depth analysis of the accident may disclose that the true cause was selection of the wrong machine for the job, or improper planning of the way the job should be done. In a sense, this is still operator error; he should recognize the limitations of the machine he is using and if instructed to do anything that is beyond the safe operating range of the machine, he should not attempt it. Whether or not a worker will question his supervisor's instructions can be an important safety consideration. Try to get across to foremen that such questions are to be welcomed and not resented. If a supervisor issues instructions to do an unsafe job and someone gets hurt, both he and the machine operator must share the blame.

Safety hazards while equipment is being repaired are increased by the possibility that while one system is being repaired a defect in another system may cause injury. For example, a small fuel leak might be present for many days and never be noticed while a machine was being used, but the same small leak could be responsible for a fire in a repair shop. It is best to assume that when one thing is out of order, more than one is probably in need of attention. Don't rely upon the hydraulic system to support a scraper blade or bucket while repairing some other system. Anticipate trouble with wiring while working on fuel lines; disturbing a frayed wire may cause a spark and ignite fuel. A pinhole leak in a high pressure hose or tube may emit a fine stream of fluid that can penetrate the skin. This characteristic of high pressure fluids is useful as well as dangerous; medical injections without the use of hypodermic needles are an established procedure, but who wants an injection of hydraulic oil?

Safety regulations should prohibit machine repair work while only one worker is in the shop. If anything goes wrong, he may urgently need help and be unable to summon it himself. Even when several people are available, it is dangerous to work on a machine without anyone at the controls while the engine is running. Remember also that carbon monoxide gas is poisonous and odorless.

Inexperienced workers are apt to suffer hernias or back injuries when they try to handle heavy parts without proper equipment. When field repairs are attempted, be sure the proper tools are available before starting the job; don't improvise. Tire work is more difficult and dangerous than many mechanics realize. Tire specialists know when it is important to use cages, chains, or long reach tools to inflate tires without endangering themselves. The compressed air in a large tire can do a great deal of damage if anything lets go.

I have found that when something is wrong with a machine and customary repair methods are ineffective, it is best to call the president of the company that made the machine. If it is a large company, the local librarian may have a directory of corporations that lists the names of the principal officers of the company. By all means start by calling someone high enough on the corporate ladder; it is far easier to get action when the person called refers your call to one of his assistants than to induce someone in a subordinate position to request approval for emergency action. Keep a record of the names and positions of all people to whom you speak, the time and date, and what was said. People are more apt to act responsibly if they realize they may be quoted. It has been my observation that people who have real authority to act are the ones who get things done without quibbling about who will pay the bill for a special trip to the job if necessary. The designer of a piece of equipment needs to know firsthand about troubles; he may even come to the job himself if he never heard of the particular type of trouble before. On the other hand, if you call and ask for someone in the maintenance department and happen to get a clerk on the phone, you are apt to be asked to fill out forms, write letters, call back, or do anything the person can think of to get you off his phone without having to become involved in a problem he does not know how to handle.

5.3 Handling and Driving Piles

It should be self-evident that it is necessary to mate the pile driving equipment with a crane or derrick capable of handling the total weight of the hammer, pile, and leads at every necessary location, but it is possible to overlook the fact that piles are generally used where soil conditions are unstable, and the rating of the crane under actual field conditions may be considerably less than its maximum under ideal conditions. When batter piles are to be driven, the angle of the pile from the vertical also affects the loading on the crane. When a crane needs to be operated near its maximum capacity, the effects of these variables should be evaluated by an engineer and not left to the judgment of the operator.

Accidents can happen while piles are being unloaded from the truck, car, or barge on which they were shipped to the site. Unless given specific loading instructions, the shipper's primary interest is in securing the load so it will arrive safely at its destination. Whether it can be unloaded safely or not is of little or no interest to him if the purchaser has to do that work. Plan the handling of the piles with safety in mind. Keep workers clear of spots where they could be injured, such as alongside of a car that is about to be unloaded.

While working near or over water, do not anticipate that workers will know how to swim or that in case of an accident they will be in physical condition to do so. OSHA requires that life jackets, ring buoys, and a lifesaving skiff all be at hand when work is being done over water. In winter, consider also the danger of death being caused by exposure if a worker falls into the water and cannot be rescued for even a few minutes. (See also Section 3.2 on worker' compensation insurance.)

When wood piles are driven through water and the butt ends have to be trimmed off to a specified height, keep the short pieces under control so they do not become a navigational hazard. Remember that some of the best tropical woods used for piling are so heavy that they barely float.

In Chapter 4, Section 4.4, I recorded some thoughts about avoiding damage to other structures while blasting. These comments also apply to vibrations caused by pile driving. Air compressors can contribute to the vibration problem. The courts have ruled that damage to adjacent structures is the responsibility of the contractor. In his Wiley Practical Construction Guide, *Construction Failure,* Jacob Feld has recorded some interesting case histories, including one example that involved destruction of buildings fronting 400 feet on a street where the pavement settled as much as 16 inches. (See page 58 of Dr. Feld's book.)

Heavy impact hammers that operate at relatively slow speeds produce more noticeable vibrations than high speed vibratory hammers, but I would hesitate to assume that the high speed equipment will eliminate vibration damage. Resonant effects are difficult to predict. Sometimes it may be wise to avoid vibration problems by substituting piers and caissons for piles. A change of this nature may appropriately be recommended by a contractor if his inspection of adjacent structures reveals conditions that were apparently unknown or inadequately evaluated at the time the foundations were planned. If the contractor is to be held responsible for damage, it is unreasonable to require him to do work he is confident will cause damage in spite of all precautions he may be able to take.

5.4 Compressed Air

If air hammers or drills will be used, a short tool box safety meeting is appropriate before the compressors are started. The discussion leader may want to review material previously covered about noise, or to bring up the subject at this meeting (see Chapter 3, Section 3.3.3).

Workers should know how to inspect tools every time they start work. Hoses must be in good condition and securely attached. Bits must be secured to impact tools so they cannot become accidentally detached even if the tool is accidentally started while it is lying on the ground. Hoses must not be used to lower or lift tools; provide ropes for that purpose.

Nobody likes to use respirators, but if drills or other air actuated tools produce dust, it must either be wet down at the source or respirators must be worn. Under some conditions, other means of dust control may be possible, but under field conditions they are seldom feasible unless a deep excavation must be ventilated and dust control can be achieved at the same time.

In cold weather when battery failure may make it necessary to manually crank a small engine, workers may not know how to do it safely. Emphasize the importance of lifting on the crank and of having fingers and thumb on the same side of the handle. The spark control may have to be manually retarded to minimize the danger of kick-back. When starting fluids are used, be sure to enforce the no-smoking rules.

Safety precautions when handling gasoline are so important that repetition of the topic at several safety meetings is warranted. Always use safety cans; keep them capped. Make contact between the can and the gas tank before beginning to pour and maintain contact; a spark can be created as a result of the flow of liquid if there is no metallic contact between can and tank. Label containers and keep them in a safe place where they will not get knocked over, exposed to heat from the engine exhaust, or used for any but their intended purposes.

If operations make it possible to assign one worker to take full responsibility for starting, stopping, adjusting, and maintaining air compressors, it is desirable to do so. If any of several workers may have to do this work, be sure all are fully qualified, and are aware of the dangers inherent in the use of compressed gases of any kind. Never allow anyone to try to fix a relief valve; install a new one. In anticipation of possible trouble of this kind, I suggest keeping a spare relief valve on hand so nobody will be forced to choose between safe operations and shutting down the job until a new valve is obtained.

Post safety rules governing use of compressed air. Do not permit high pressure air to be used to blow dust off clothing and under no circumstances allow any horseplay. If hand signals are needed, check to be sure workers understand them; just posting a set of pictures is not enough. Who is going to go look at a picture when it is necessary to signal someone?

For specific OSHA instructions, see 29CFR 1926.302(b).

5.5 Shoring and Sloping of Trenches and Excavations

At a Boston subway project, part of the timber and steel shoring of an excavation gave way during the lunch break. One worker died and another, who was buried under 10 feet of sand, was rescued. Twenty others narrowly escaped; they had left the excavation only a few minutes before the accident.

Even though scores of workers are killed or injured every year because excavations were improperly sloped or shored, their stories seldom make headlines. When a trench caves in, no high technology is likely to have been involved; what can the reporter say? He may be able to find a human interest angle so his editor will feature the story, but his job is far easier when a boy is trapped in a cave or a plane cannot land because it has lost a wheel. Nonetheless, to the people involved, the loss is just as important as it would be if there were headlines in every newspaper.

Sometimes the foreman is to blame for the accident. One grudgingly ordered a trench to be shored only when a safety inspector ordered that he do so. Later, when an unshored extension of the same trench collapsed, the foreman himself was injured. This type of bullheaded person cannot be taught; he should never be in a supervisory position.

Sometimes incentive payment plans encourage unsafe practices. If payment is made on the basis of the number of feet of pipe laid and there is no allowance made for variations of soil conditions or recognition that the cost of shoring will depend upon the depth of the excavation at different places on the site, it is only human to expect that chances will be taken.

There is more skill required in properly digging and shoring a trench than meets the eye. A simple straight cut in uniform soil may not be too hard to make, but where the trench goes from one type of formation into another, the geology of the site becomes an important consideration. Add to this the complicating factor of intersecting cuts, curves, variations of depth, and groundwater conditions and the foreman's job is no longer an easy one. Some excavations are also considered to be confined spaces; oxygen deficiency and toxic or explosive gas pockets may be encountered (see Chapter 7, Section 7.7).

OSHA has some minimum safety regulations. If trenches are four feet or more deep, a ladder or other means of exit from the trench must be provided within 25 feet of each worker. Figure 5.1 and Table 5.1 are reproduced from 29CFR 1926, subpart P, paragraphs 1926.650 to 1926.653. They apply to trenches five feet or more in depth and require shoring or adequate sloping if there is reason to anticipate hazardous ground movement. Still, considerable judgment is needed to apply the tables. Note, for example, that the soil conditions are described, but it is left to the people on the job to decide which description fits the soil being excavated. I recommend that safety engineers read the reference cited, and if soil con-

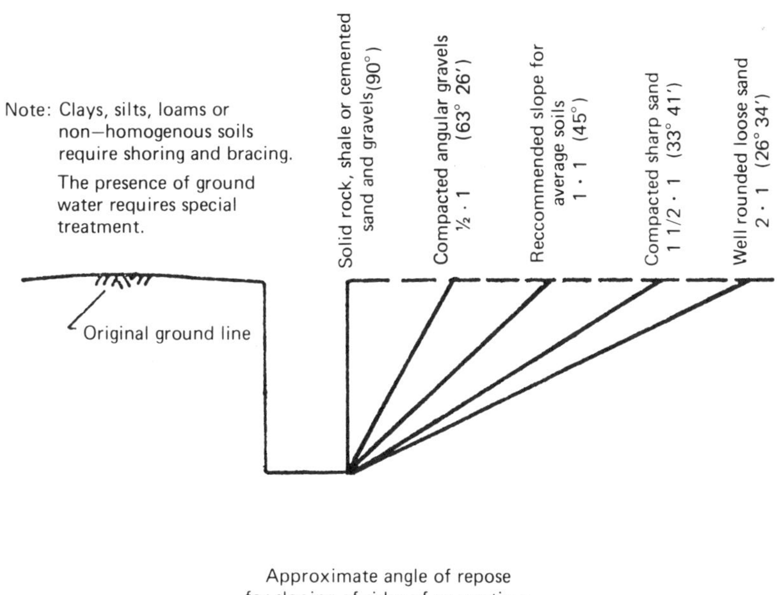

Figure 5.1 Approximate angle of repose for sloping sides of excavations. (Reproduced from OSHA regulations, 29CFR 1926, Subpart P (Table P1), and is applicable to trenches and other excavations over five feet deep.)

TABLE 5.1 Trench Shoring—Minimum requirements

		Size and Spacing of Members			
		Uprights		Stringers	
Depth of Trench	Kind or Condition of Earth	Minimum Dimension	Maximum Spacing	Minimum Dimension	Maximum Spacing
Feet		Inches	Feet	Inches	Feet
5 to 10	Hard, compact	3×4 or 2×6	6		
	Likely to crack	3×4 or 2×6	3	4×6	4
	Soft, sandy, or filled	3×4 or 2×6	Close sheeting	4×6	4
	Hydrostatic pressure	3×4 or 2×6	Close sheeting	6×8	4
10 to 15	Hard	3×4 or 2×6	4	4×6	4
	Likely to crack	3×4 or 2×6	2	4×6	4
	Soft, sandy, or filled	3×4 or 2×6	Close sheeting	4×6	4
	Hydrostatic pressure	3×6	Close sheeting	8×10	4
15 to 20	All kinds or conditions	3×6	Close sheeting	4×12	4
Over 20	All kinds or conditions	3×6	Close sheeting	6×8	4

[a]Trench jacks may be used in lieu of, or in combination with, cross braces.
Shoring is not required in solid rock, hard shale, or hard slag.
Where desirable, steel sheet piling and bracing of equal strength may be substituted for wood.

ditions are difficult, consultation with specialists in soil mechanics may be advisable.

Note that Table 5.1 permits vertical trench walls without shoring if the cut is through "solid" rock, but what is "solid"? Most formations are fractured, or become so as a result of blasting. In many locations, sedimentary rocks such as sandstone that were originally laid down as horizontal layers have been deformed so that the bedding is no longer horizontal. Planes of weakness can permit slippage to occur. In formations where slippage has occurred under conditions of heat and pressure, the rock faces may have become polished or "slickensided" and if a structure of this type is inclined, dangerous slides can occur whenever a trench or excavation of any kind is made through it. Many museums have specimens of slickensided serpentine from the historic Tillie Foster mine in Brewster, New York, where rock slides caused so many fatalities that all mining had to be stopped.

Judgment must also be used when selecting appropriate slope angles

Size and Spacing of Members						
Cross Braces[a]						
Width of Trench					Maximum Spacing	
Up to 3 Feet	3 to 6 Feet	6 to 9 Feet	9 to 12 Feet	12 to 15 Feet	Vertical	Horizontal
Inches	Inches	Inches	Inches	Inches	Feet	Feet
2×6	4×4	4×6	6×6	6×8	4	6
2×6	4×4	4×6	6×6	6×8	4	6
4×4	4×6	6×6	6×8	8×8	4	6
4×4	4×6	6×6	6×8	8×8	4	6
4×4	4×6	6×6	6×8	8×8	4	6
4×4	4×6	6×6	6×8	8×8	4[b]	6
4×6	6×6	6×8	8×8	8×10	4	6
4×6	6×6	6×8	8×8	8×10	4	6
4×12	6×8	8×8	8×10	10×10	4	6
4×12	8×8	8×10	10×10	10×12	4	6

[b]This dimension does not appear in source, as of 1982.
Source: Reproduced from OSHA regulations, 29CFR 1926, Subpart P (Table P2) and is applicable to trenches five feet or more in depth.

for various soils. If banks are to remain as a permanent feature of the site, consider the effects of varying moisture content of the soil and when piling material removed from a trench, be sure it is sloped adequately to prevent sliding. Be particularly cautious when clay is mixed with other types of soil in the same pile.

The Culebra Cut, which carries the Panama Canal through the continental divide, is a classic example of how the size of an excavation can affect the stability of the sides of a cut or excavation. Soil or rock formations that would be stable for shallow excavations are apt to be unstable when deeper cuts are made. At Panama, a board of consulting engineers estimated that a suitable typical slope for the walls of the cut would be about 56 degrees from the horizontal and that vertical walls would be permissible where basalt or granite formations were penetrated. But the rock walls broke, and even forced material up into the bottom of the cut. One large earth slide continued to move until the rise was only one foot in ten. Some slides were worse in the dry season than in the rainy season. In all, over

250 acres of land that were expected to remain in place slid into the excavations, including 75 acres of the town of Culebra on which several hotels had been built. In a single year slides destroyed 200 miles of railroad track and a number of steam shovels were buried. The original estimate of 54 million cubic yards of material to be excavated was revised to 70 million and the final actual total was 100 million cubic yards.

Table 5.1 summarizes OSHA's minimum shoring requirements for trenches. Trench shields may be used instead of shoring.

At all times, be alert to the possible effects of vibration or pressure caused by traffic, pile driving, movement of heavy equipment near the excavation, or the weight of material removed from the cut and piled near it. In general, it is good practice to keep piles of earth back from the edge of the cut by at least half the depth of the cut. Watch particularly for rocks that could roll into the trench. See Section 3.3.7 regarding earthquakes, including minor tremors.

The public must be protected during this phase of the construction work by the use of barricades, warning signs, and lights. In addition, watchmen are often needed, particularly if children may be tempted to play nearby. Backfill trenches as soon as possible.

When removing shores, follow a planned sequence so that workers are safe during this critical part of the job. Leave loosened cross-bracing in place until workers are out of the trench and then use ropes or mechanical equipment to remove the shoring.

For additional comments on trenching problems, see *Construction Guide for Soils and Foundations* by Gordon A. Fletcher and Vernon A Smoots, another volume in this Wiley Series of Practical Construction Guides.

5.6 Laying Pipe in Trenches

Safe handling of pipe begins with planning for unloading and stacking it before it is delivered to the job. Get skids and whatever barriers will be needed. Keep the storage area clear and ready so that makeshift methods will not have to be used and to save the cost of rehandling. These precautions are of course applicable to all materials, but pipe's tendency to roll and the importance of avoiding damage to protective coatings and entry of dirt into the open ends make preparation for receiving shipments particularly important.

Be sure piles of pipe are adequately blocked; keep everyone clear of the sides of the pile until it has been made safe. Work from the ends of the pile when moving pieces of pipe.

If individual lengths of pipe are strung out along a trench, it looks very orderly to have them parallel to the cut. But it is far safer to lay them at an angle to it; if a piece begins to roll, it is much less likely to fall into the trench.

For sizes of pipe that will be lifted by hand, use of hard wood pipe sticks helps avoid hand injuries. If slings are used, check them frequently for wear; choker slings used repeatedly for moving the same size pipe tend to wear unevenly. Keep the pins of shackles clear of the running part of a rope that could loosen them.

If it is necessary to roll heavy pipe down skids, plan the work so that friction between the pipe and ropes or between the rope and a fixed support will take the strain rather than the workers. Never allow a worker to wrap a rope around his arm or sling it behind his back to help him hold it securely. Note, for example, that if the fixed end of a rope passes between the skids, under the pipe, and over it to the worker, he has to resist the rolling weight without help from any frictional resistance between the rope and the pipe. On the other hand, if the rope is run from a fixed support over the pipe and around it to the worker, the rope must slide as the pipe rolls and friction between the pipe and the rope can take most of the stress off the worker. Similarly, snubbing the free end of the rope around a post or tree makes it very easy to hold. But regardless of how the work is planned, keep everyone clear in case the pipe breaks loose and rolls free of all restraint.

Accidents have happened when pipelines have been supported by timbers placed across trenches in preparation for lowering into the cuts. If one support lets go, it may initiate a domino effect, so that the next and then the rest of the timbers snap one after the other until the entire run of pipe has fallen. Workers have been trapped at some distance from the initial break as a result of this progressive support failure.

When welding is required, protect other workers from glare. If a laser is used to align the pipe, be particularly cautious about the possibility of reflection of the beam from any shiny surface into the eyes of a worker. Mica in rocks, broken glass, and metal scrap should be kept out of the laser beam.

If wood shoring is used while welding is being done, be prepared to put out fires before they get out of hand.

Backfilling must not be done until it has been confirmed that all workers are out of the excavation, and a watcher is needed to help a bulldozer operator if he has any difficulty seeing exactly what he is doing.

5.7 Water Supplies

In a book about safety, it is not necessary to discuss the contractor's problem of getting an adequate amount of water for construction workers to drink and for uses incidental to the work itself, such as mixing concrete, fire protection, and cleanup, but I do want to point out some safety hazards to be avoided.

If water is to be obtained from wells, in some localities subsidence of the earth or collapse of sinkholes may be problems. Local well drillers and the state geologists may be able to predict problems and suggest precautionary measures. There have been cases where other users of water from wells in the same general vicinity have found that drilling a large well has lowered the water table and caused their wells to run dry. Even if this claim is not justified, an injunction may be obtained and work held up. Again, well drillers and geologists can be helpful.

Water wells near the sea shore may turn brackish if overpumped and, if this occurs, it may not be possible to continue to use the well, even at a reduced pumping rate. It may take years before the underground water conditions return to normal after being disturbed in this way. It is therefore advisable to make frequent tests and to stop pumping at the first indication that sea water intrusion may be occurring. Of course, the use of salty water for mixing concrete is not to be recommended for structural reasons.

Use of water from streams or springs near wetlands that are protected by law from being drained or filled in can draw fire from conservation groups and work may be stopped by order of the Environmental Protection Agency. Similar action can be expected if heavy soil erosion follows clearing work, and mud and construction wastes endanger wildlife.

At the time that the water supply is being arranged, there may be little or no visible need for water for fire fighting, but there may be a time when a temporary water supply may be needed. It is a responsibility of the safety engineer to be sure that there is no interruption of the availability of enough water for this purpose, even if it is necessary to bring in portable tanks.

5.7.1 Potable Water

When a project has been completed and people are using water regularly throughout the building, any sudden discoloration, unpleasant taste, or odor will result in a complaint to the manager. He does not have to monitor the water supply; others will do it for him.

But at a construction site, if it is near the end of a water main or if work is being done on the supply lines, sediment and organic matter in the pipes may be disturbed. If the water is put directly into the drinking water lines, they will all have to be flushed out before the water runs clear again, which is a nuisance even when it is not a safety hazard. If very dirty water goes through a hose into a concrete mixer, the contamination may not be noticed and it may affect the mix. Someone should make arrangements with the water supply department or water company, so that prior notice will be given before work is done that will disturb sediments. At the same time, keeping an eye on the quality of the water before every important use is made of it can avoid trouble.

Good potable water can be contaminated on the job. Hoses lying on the ground can get mud in them. If an eductor is used to drain a pit, the discharge water may look clean enough to tempt a thirsty worker to grab a quick drink without paying attention to the connections. Warn workers never to drink from a hose; provide suitable drinking fountains where safe water is available.

Siphoning of contaminated water into the water supply is very often possible. Whenever a water supply pipe is drained, a partial vacuum is created in it and, if a hose connected to the line ends below the water level in a drum or wash trough, the dirty water will be siphoned into the pipe until air entry breaks the siphon. When the pipe is refilled, water passing through it may be dangerous to use. Sanitary codes require vacuum breakers at all hose connections, but even in completed buildings they are often missing. Safety requires that they be used during construction.

If there is a fire in the neighborhood, the fire department may pump so much water from the hydrants that the city water pressure is reduced to practically zero, or a partial vacuum may even occur in the main. When this happens, any condition that is a potential siphon will immediately begin to contaminate the water unless a reliable backflow prevention device has been installed between the main and the water lines on the site.

Immediately after a power outage, low water pressure may be expected if nearby buildings have house pumps. As soon as power is restored, all the pumps will probably begin to run at the same time, causing an abnormally heavy demand for water from the main. Again, siphonage may occur.

Keep nonpotable and potable water supplies separate; don't rely on check valves to keep water from one supply out of the other. Avoid using a potable water hose to prime a pump that handles dirty water. Cross-connections between the drinking water and contaminated water are a common hazard.

Information on the foregoing problems may be obtained from most local health departments, or from the American Water Works Association, 6666 West Quincy Ave., Denver, Colorado 80235.

See also *Water and Piping Problems in Large and Small Buildings, A Troubleshooter's Guide* by H.L. Shuldener and J.B. Fullman, John Wiley and Sons.

It may be more difficult to keep the drinking water lines on the site in safe condition than to provide portable drinking water dispensers, cups, and containers for used cups. If this approach is used, OSHA requires that signs be placed at all nonpotable water outlets. See 29 CFR 1926.51 for details.

5.7.2 Water for Construction

Let me repeat an important warning. Don't get caught without enough water to fight a fire.

If a steam boiler is used, boiler water treatment is advisable. But don't just get a supply of anything labeled boiler water compound and feed it into the boiler. Have the water analyzed by a reputable water treatment firm and get the right treatment chemical for the job. It is necessary to consider the composition of the water, the type of boiler, the steaming rate, pressure, and use to be made of the steam. Nobody needs trouble with boiler operation on the job, and if the treatment is not right the boiler may scale up; tubes may fail. Or the boiler water may carry over with the steam, causing the boiler's low water cut-out controls to stop the burner operation. A boiler on a construction job is always a potential safety hazard, and should be operated by a qualified worker.

Whenever steam is used to thaw frozen aggregate or to heat water for mixing concrete, I recommend that the boiler be blown down often enough to avoid foaming, and that the sudden use of steam at a rate in excess of the boiler rating be avoided; it causes priming of the boiler. This is bad for the boiler and could damage the concrete. Proprietary boiler compounds contain many different chemicals, but the actual ingredients are seldom disclosed. The label must say something about the ingredients, but generic terms such as "dispersants" can cover a variety of different chemicals. Some organic materials such as sugar are known to inhibit the setting of concrete, and other materials may have a similar effect. Boiler water treatment compounds commonly include tannin or polymers used as dispersants to avoid the formation of hard boiler scale. They could conceivably interfere with the setting action of concrete. Why introduce an unknown variable that could weaken the structure? Operating the boiler within its rating will keep the boiler compound in the boiler and out of the concrete.

5.7.3 Corrosion and deterioration of old lines

Cast iron pipe, valves, and fittings may be greatly weakened by a form of corrosion known as graphitization. Iron castings contain flakes of graphite intimately mixed with the metallic iron, and corrosive action often removes the iron without visibly affecting the size or shape of the original article. The remaining material has very little strength and can be easily cut with a knife. This form of corrosion may occur from the inside or outside of the pipe; it is common where the ground water is brackish. Sometimes the action is spotty; one section may be severely weakened while adjacent sections are relatively intact.

Concrete pipe may also be very much weakened by the action of hydrogen sulfide released when sulfate reducing bacteria are present. This is to be expected in sewers, but it may also be present in potable water lines.

When an excavation is made to expose an old line, or work is done near one, there is always the possibility that it may suddenly break. Be particularly careful when the pipe may be under unusual stress as a result of temperature changes. Water lines under pressure should be shut off whenever there is a question about their strength. Don't plan to shut off the water after a break occurs; if the pipe is in bad condition the valves are likely to be in similar shape, and it will be impossible to stop the flow.

5.8 Electric Power

On most jobs, the electricians will know what power is available; throughout the United States the standard frequency is 60 Hz and three phase supplies are usual. On new construction work, the power company representative will probably point out any unusual characteristics of the system. But contractors who do any incidental remodeling work or have occasion to tie into private power supplies should be a little cautious. In the Greenwich Village area in New York City, I once ran into direct current and alternating current outlets in the same building. At a meeting one evening, I had a similar experience elsewhere, and burned out a projector motor by plugging it into direct current. Shipboard electrical systems may be D.C. but some have A.C. outlets also. For special uses, industrial or laboratory buildings may have some 400 Hz power lines, and where slow speed motors are used, plants with their own generators may have some 25 Hz lines. In anticipation of installing some equipment in a plant in Philadelphia I shipped a few three phase motors to the job, only to find that the plant had two phase power, supplied by the local utility. In foreign countries it is of course necessary to check carefully and constantly; 50 Hz power will not usually do permanent damage to 60 Hz equipment, but it will slow down most motors, and I would not want to generalize about its effect on other electrical equipment. Whereas 110 to 120 volt outlets are common in the United States 220 volt power is equally common abroad, as many tourists have learned the hard way.

It is not common for serious accidents to result from differences such as those mentioned, but costs go up when equipment must be returned or reordered. The safety engineer must be responsible for electrical grounding of all equipment or the provision of ground fault circuit interruptors, and so he might as well check for unusual power characteristics as well.

The safety engineer should take particular care to become familiar with the location of all overhead power lines on the site and at any location where cranes or other large equipment for which he may be responsible will be moving. See Section 6.2.2 for further comments on crane accidents and minimum clearances.

The following comments on electric system safety apply to the whole construction period. Some contractors will bring portable generators to the site before power is available from utility lines, and the electrical contractor or safety engineer must be cognizant of the hazards involved in the use of such equipment. Fuel must be kept in proper containers, and fire from exhaust systems may be a hazard. For small generators, usually under about five horsepower, it may not be necessary to ground the wiring. Check local codes.

As soon as temporary power lines are needed, the electrical contractor should do all the wiring. He or someone else must also be responsible for checking the condition of all extension cords and tools. Don't overlook this responsibility; be sure someone does this work regularly.

Keep the electrical contractor on the ball, so that other trades will not be tempted to jury rig their own temporary power lines. Workers who have a little knowledge of wiring can get themselves and others into trouble. I once checked an outlet box to see whether it was securely attached and, as soon as I touched it, a loose connection inside it shorted and burned a hole through the metal. The burn was not serious, but was painful and could have started a fire. It was probably caused by a wire connector that was loose, allowing the joint to overheat and melt the plastic insulation.

Handymen electricians seem to consider that anything they do is satisfactory as long as they get the light to light or the portable tool to run. They sometimes interchange the white neutral and green grounding wires, or use any handy pipe as the ground or neutral connection. If they put in a switch or fuse, it is likely to be found in the neutral wire. I have even seen bare wires hooked together as a substitute for a switch.

There are lots of ways to get into trouble; sometimes I wonder how many wrong ways there are to do a simple job. For example, some small transformers for stepping down 120 volt power to 12 volts have 2 primary leads to be connected to the 120 volt line, and 2 secondary leads from which the 12 volt power is to be drawn. Of course, there is some grounding connection also. The instruction sheet will give the key to the color coding or location of the leads, if they are not marked clearly. But, of course, it is always a possibility that someone will get the 120 volt wiring connected to the 12 volt leads and, until something burns up, there will be 1200 volts across the 120 volt leads. People tinker with oil burner controls on the assumption that they are wired for 12 or 24 volts, forgetting that the ignition transformer probably steps up the 120 volt supply to 10,000 volts or more.

Severe flash and burn injuries can be caused by shorting batteries; the high capacity ones used in electric fork trucks and similar mobile equipment are particularly dangerous. Dropping a tool across the terminals can cause a lot of fireworks. Hydrogen gas released during charging operations can cause the atmosphere to be explosive. Prohibit smoking where batteries are being charged.

Electricians often set a bad example for other workers by touching wires they know to be hot. The uninformed worker then assumes he may safely touch a frayed wire or pick up a defective tool under conditions that a trained electrician would know could be fatal. It is not just the circuit voltage that determines the risk of injury; it is the amount of current flow through the body and the portion of the body through which the current passes that are critical. Dry skin has fairly high resistance to low voltage power. That is why we often see an electrician touching a 120 volt wire with two dry fingers of the same hand without visible discomfort. His skin resistance is probably well in excess of 100,000 ohms, and so the current flow would be roughly one milliampere. This current is about enough to be felt as a mild shock. However, if the skin is wet or punctured by a sharp spur on the energized conductor, skin and body resistance drops rapidly and the current increases immediately. If the current reaches about 10 to 15 milliamperes the person may be unable to let go, or may have violent muscular contractions that can cause a fall. It takes less than 100 milliamperes to cause death and, of course this is only one tenth of an ampere or just about enough current to light a pilot light. If current flows through the heart or other vital organs, breathing may stop, normal heart action may cease, or other internal injuries may occur.

The chances of surviving a severe shock are slim, but first aid to restore heart action and breathing may be effective if provided promptly.

Workers should be warned that it is difficult to predict whether touching any hot wire will or will not cause a severe shock, and that the only safe action is to leave all electrical work to people who are trained to do it safely.

When journeymen electricians provide temporary power and use reasonable care, there should be no reason for them to be injured by shock, but when high voltage power or heavy currents are involved, even licensed electricians may need special training. All others must be kept away from such hazards. Safety engineers who do not consider themselves adequately qualified to conduct safety meetings and training sessions about high voltage or high current work should arrange for specialists to be present. Note that even equipment that runs off a low voltage power supply may have very high voltage components inside the device. No worker should be assumed to be qualified to work on equipment circuits unless his qualifications have been confirmed in some way.

Safety engineers should be certain that the electrical contractor supplies his employees with proper tools and equipment. Don't allow metal ladders to be used. Be sure instruments and insulating gloves are supplied when needed. Hard hats must be made of insulating material, designed for electricians.

Include electricians in general safety meetings. Most of their work is done before circuits are energized, and they fall off ladders, injure themselves handling materials, and are struck by falling objects about as often as any other workers.

The safety engineer and electrical contractor should review OSHA regulations, the National Electric Code, and the local code together before work begins. Consider particularly the rules governing grounding of tools and equipment, and the use of ground fault circuit interrupters. You may find that the ground fault circuit interrupters are optional in your locality, or their use may be mandatory. Be sure the grounding connections have low enough resistance to permit circuit breakers or fuses to interrupt the current in case a hot wire is grounded. Driven rod grounds may meet resistance requirements for small projects but be inadequate if heavy currents are used at the job.

Misunderstandings regarding the National Electric Code are not uncommon. It is a compilation of recommended practices prepared by the

National Fire Protection Association, but it is not a binding regulation until it has been adopted by a regulatory body having local jurisdiction. Some cities recognize the latest revision of the National Electric Code; others specify that a certain edition of the Code be followed. Always assume that the edition mentioned in the local regulations must be followed unless you are sure the local authorities intend that the latest edition be used. I note that the 1982 version of OSHA regulations, 29CFR 1926.400 refers to the 1971 edition of the National Electric Code, even though two revisions of that Code have appeared since 1971. Regulatory bodies often hesitate to accept rule changes without reviewing them, and may withhold approval for many years.

OSHA regulations formalize many commonsense rules with which it is unlikely anyone will quarrel, but it is a good idea to look over 29 CFR 1926.400 to 1926.405 as a checklist. Note that if the assured equipment grounding conductor program option is to be followed instead of the use of ground-fault circuit interrupters, test records are required and every item must be identified. This can be bothersome, but keeping the ground-fault circuit interrupters from cutting out can also be a nuisance. An even greater problem is, of course, keeping workers from nipping off the grounding prongs, using two conductor cords, and doing all the other things you and they know are unsafe but which it seems at times almost impossible to prevent.

Workers have been electrocuted when cranes made contact with an overhead power line (see Chapter 6, Section 6.2.2).

Electricians will probably be among the first workers present at an accident involving contact with live wires, and they should be trained in cardiopulmonary resuscitation (CPR) as well as basic first aid.

5.9 Other Utilities

In some places steam, high temperature hot water or chilled water are offered in addition to the usual electric, water, and gas services, but before any service is turned on the utility will probably have an inspector visit the job to be sure safety precautions have been taken. However, on college campus projects, in industrial plants, or wherever the service is provided from the owner's power house or boiler room, close coordination with the operating staff is essential to be sure pressure is not turned on until the construction work has progressed to the proper stage.

5.10 Placing Concrete Footings and Foundation Walls

Good construction practices and safety go hand in hand when transit mix trucks and cranes are moving about on a site. To keep equipment from getting mired down, tipping over, or skidding is as much a production problem as a concern of the safety engineer. Accidents are most apt to happen when something delays the work and to make up time the pouring rate is speeded up or form work is rushed.

Every job is different, and it is not possible to make recommendations that apply to all situations, but to expect a foreman to supervise his men, direct truck movements, anticipate emergencies, and still watch every small detail that could cause an accident is probably asking too much of him.

Placing concrete is often the culmination of a lot of preparatory work, the big day when it is essential that everything go according to plan. It is therefore important that the people in charge of the activity be relieved of all minor duties so they can watch for possible troubles and move quickly if anything goes wrong. The safety engineer may need some temporary help so that he is constantly available. If there are filled areas where the earth is soft, he may need a flagman to keep drivers who have never been on the site away from danger spots. Someone may be needed to be certain that wheel chocks are in place or to watch for any shifting of runway supports if concrete buggies are to be used. An extra pair of eyes may be needed to be certain that everyone is out of the way before concrete is poured and that a form failure could not cause injuries.

Foresight could probably have saved two lives and injury of 10 other workmen at a small power dam project along the Ohio River. A pigtail anchor failed and a vertical strongback was left unsupported. This caused six more anchors to snap, one after the other as the form collapsed. One man fell from a scaffold; another was working near the bottom of the form. Both were buried in 25 cubic yards of concrete and killed.

Pouring footings may be the first operation at the job for which a crane is required; it may also be the first time the crane operator and his signalman have ever worked together. Standard signals should be used at all times; someone should check to be sure they are doing so. Under field conditions the maximum load a crane should be expected to handle at the farthest point from its cab may be very considerably less than the manufacturer's rating, based on ideal conditions. Consider whether a timber mat is needed to assure a secure base for the crane; don't put the operator in a position where he must take chances.

Extra workers may be needed during concrete placing operations, and untrained workers are most apt to suffer muscle strains and back injuries. Sports do not necessarily develop the muscles that are needed on the job, and a young man who seems strong and healthy may not be in condition to do the work of a 50 year old who weighs 50 pounds less. I recall watching a man in his sixties handling 600 pound drums, moving them from a storage area onto his trailer truck with only a hand truck and a dock board. Within far less time than most of the younger men would have taken to do the job, he moved 40 or 50 drums without any apparent signs of fatigue. He did it often, and had the knack of keeping everything in perfect balance

Figure 5.2 Laborer vibrating concrete on a 500 foot hyperbolic cooling tower shell placement. He is standing on top of the form about 6 feet above the scaffolds on either side of the form. Note attachment of his safety lanyard to the steel cable supported by the rebars. (Photo courtesy Philadelphia Electric Co.)

so that his spare 140 pound body was never overstressed. Youngsters on the job for the first time are apt to think they know how to handle a wheelbarrow or shovel and before they have done even half a day's work they have put more strain on certain muscles than they can absorb without injury. I realize that it is hard to avoid pushing a worker who is goofing off, but if he is unable to do a day's work and pulls a muscle, the medical expenses and compensation payments to which he may be entitled can far exceed the value of the work he should have done while he was leaning on his shovel. It is important to distinguish between the employee who is out of shape and needs to get the hang of the job from the one who thinks he is under no particular obligation to give his boss a day's work for a day's pay.

When reenforcing rods are left protruding above concrete, be particularly vigilant against the possibility that someone may fall on one of them. The need for special attention to the safety of workers on scaffolding is obvious, but a serious injury can also result if a worker stumbles at ground level and falls on even a short rod. In cold weather, if concrete is covered with salt hay, stumbling hazards and rods may both be hard to see, and some type of marker may be needed.

OSHA regulations cover a number of topics that I have not mentioned. There are some specific rules, such as the requirement that powered troweling machines be equipped with a switch that will shut off the power whenever the operator removes his hands from the handles. I see no reason to repeat all these rules; they should be checked, however, by the safety engineer before concrete work begins. See 29CFR 1926.700 through 1926.702.

6

Erecting and Closing In the Structure

6.1 Safety Program Review and Revisions

In Sections 2.1 and 2.2 of Chapter 2, I made some suggestions for safety engineers concerning the planning of a safety program and the motivation of workers. It was appropriate to do so in anticipation of problems because a successful program must be planned. I also mentioned in Section 2.4 that programs must often be changed to adapt them to unanticipated conditions.

The safety problems during erection of a structure are obviously different from those that predominate during excavation and the preparation of foundations. Injuries caused by falls, material handling, and fire are apt to become more frequent. Special hazards are associated with roofing; structural failure may occur; or there may be conditions that will lead to subsequent failure. Changes in the weather may have a significant effect on the safety of workers. The safety program should be reviewed in anticipation of these changes.

I also suggested that plans be made for handling emergencies, and this section should be reviewed periodically to be sure the doctor who agreed to handle emergencies and to examine workers who may have occupational illnesses is still available. Has he handled cases referred to him in a satisfactory manner? Has he ever been difficult to reach? Check phone numbers and driving directions to reach emergency treatment facilities. Traffic patterns change, and to end up on a one-way street or to be blocked by bridge construction while taking someone to a hospital could be serious. Check first aid kits, and review the list of workers and supervisors who are qualified to give first aid.

If you are responsible for safety, what have you observed about the attitudes of your professional associates toward the safety program? Have you gotten full cooperation and support from them, or have some felt that accidents will happen regardless of what anyone can do to prevent them?

Do any supervisors seem to resent safety rules and consider them unwarranted interference with production rates or profitability?

What have you learned about the owner's attitudes? Did he seem interested in having a finished structure that will be safe? Have the architect and engineers been finding errors in shop drawings? Do they seem to take that responsibility seriously? Has anyone from OSHA been at the job? Were any citations issued? Were they justified? Have the appropriate remedial steps been taken? Do local inspectors seem to be well qualified and honest? Will anyone bear watching to be sure he does not try to evade safety rules?

Have there been any near misses or serious accidents? Have minor injuries been frequent? How does the safety record to date compare with industry averages? Do you feel that you are doing a good job as safety engineer, or should you try to have the assignment transferred to someone else? The going is likely to get rougher as the job progresses, and you should be mentally prepared for it.

Has anything newsworthy happened that could be the basis for a favorable article in the local paper? How many man-days have been worked without a lost time injury? Have any workers participated in local activities? Has any subcontractor completed his part of the project without an accident? If so, should he be recognized in a news release?

With completion of the foundations and the beginning of the erection of the structure, new workers and foremen will be coming on the job. What plans have been made to make them aware of the safety rules? Responsibilities have to be checked to see what the contract documents require each subcontractor to do about trash removal, erection of barriers and safety nets, supplying safety gear for their employees, and paying for worker's compensation insurance.

If a new foreman shows up who seems indifferent to safety, it may be necessary to go over some facts of life with him and make it clear that it is possible for a fatal accident to occur. Get him to consider that he may have to notify the worker's family, answer endless questions about why and how the accident occurred, and appear in court to defend his employer if a third party suit is filed. Some people have a blind spot when it comes to visualizing possible trouble; it helps put the safety program in perspective if possible events are discussed with them before they have to learn the hard way.

Be sure everyone on the job recognizes that skilled workers are safe workers. People who know their jobs well seldom get hurt. Wiliam G. Rapp, P.E., in his book *Construction of Structural Steel Building Frames** considers safety and skill as closely related as the left and right gloves of a pair. Rapp's instructions on safe and efficient working methods are well worth the attention of every safety engineer and steel erector.

*Another volume in this Wiley Series of Practical Construction Guides.

6.2 *Materials Handling*

Construction and mining, the two most dangerous industries, have much in common. They both involve a great deal of on-site materials handling. Manufacturing, the third most dangerous, still has materials-handling accidents in spite of greater opportunities for mechanizing routine operations. In construction work, the necessity for moving each component of the structure to its place as a separate operation makes the industry unique; there is no assembly line where workers can put parts in place as the structure moves past them. Every lift, every material handling operation, can be slightly different. It is these small differences in operations, which seem to be repetitive routines, that are important from a safety standpoint. For example, a crane may have many lifts to make, handling identical beams, but each one goes to a different spot. One lift may require extending the crane to its maximum reach while all of the others are well within its capacity. But it takes only one overload to topple a crane.

Materials-handling safety warrants the attention of every supervisor to avoid accidents with cranes, manual lifting and carrying, placing of concrete, and falls resulting from poor housekeeping. Burns are a special consideration to roofers and welders who may come in contact with molten pitch or hot metal.

Good materials handling is part of good construction management, and so I assume that the safety engineer may concentrate on those materials-handling details that may not have been adequately planned to avoid accidents.

For example, most palletized materials are piled for handling with a fork truck at the supplier's warehouse or yard. Bags are customarily cross-piled so each layer keeps the one below it from shifting, and this is usually sufficient in a warehouse or even on a truck on a public road. But when the same pallet load is moved over rough ground at a construction site or lifted by a crane, additional ties may be needed. Also, if bags extend beyond the pallet, it is not necessarily a problem in a warehouse, but if a sling cuts into a bag, the contents can spill onto workers below.

For safe and efficient handling, purchase orders should state how materials are to be packed, bundled, stacked, and tagged. If the site is extensive and deliveries are to be made to specific locations, the purchase order should state the truck loading sequence so the first items to be unloaded will be put onto the truck last. If dunnage must be provided so the material can be handled at the job, the need must be anticipated to avoid makeshift operations when the material arrives. If the material is ordered properly, it is the shipper's responsibility to straighten out receiving problems.

When palletized loads are expected, be sure to find out what type and size pallets will be used; they must be suitable for use with the materials-handling equipment available at the job. Occasionally there may be a question of the ownership of pallets; don't get caught in a situation in which you have to unload the pallets or pay an exhorbitant amount to keep them. Be sure that everybody knows who is to unload the truck and whether sidewalk delivery or inside delivery to a specific part of a building is required of the carrier. Before a shipment is expected that will require inside delivery, be sure to check the purchase order and get it corrected if it does not give complete instructions. It pays to call the carrier before difficult loads are expected to arrive, just to be sure proper information gets to the driver. Accidents happen when everybody gets angry and excited about delivery and unloading problems; some drivers are also adept at avoiding work if they think they can get away with it.

Preparations for receiving pipe or other materials that are not readily piled in a stable manner should be checked for safety. Consider the possible conseqences if a pile were to collapse. Would it fall onto a work area or traffic lane?

Structural strength of precast beams, prefabricated roof trusses, and similar items may be reduced by improper unloading or storage. For example, wood frame roof trusses are rigid when properly braced in a building, but while they are being unloaded it is easy to damage them. Defects created at that time may not be detected until a structural failure occurs. Inspection during unloading may also disclose damage that occurred in transit.

6.2.1 Vehicles

Contractors often have one or more panel trucks for general use or for specialized activities such as refrigeration equipment installation. These vehicles are part of the materials-handling activity and deserve the attention of safety engineers.

Sudden stops cannot be avoided, and loose material causes many injuries to drivers. Always have a strong barrier between the driver's seat and the space where tools or materials are carried. Truck dealers may be prepared to install them before new vehicles are delivered; see what they will do before closing the deal.

Establish a policy about riders. If two workers should ride together, provide a seat, barrier, and seat belt for each. Don't expect the second one to squat on a box next to the driver or to ride in back with the load. If a truck is not arranged to carry a second person safely, don't let him ride in it. Planning safe transportation for workers is important. Don't leave it to employees to work something out at the last minute. An accident can be a job-related case, subject to all the regulations applicable at an actual construction site.

Remember that if any hazardous materials must be handled, the Department of Transportation and all local rules apply. If a fender-bender accident happens, the vehicle may burn or explode. Local police and DOT officials are not apt to be lenient when transportation rules are evaded.

Workers who drive cars with automatic transmissions may not want to admit they do not know how to drive a manual shift vehicle, and they are unlikely to know how differently even a half ton truck handles. They get into trouble while trying to learn to drive and fight traffic at the same time. Check a worker's driving ability as carefully as his other qualifications if he is going to drive a truck. Check driver's licenses; be sure they cover the type vehicle to be driven, are valid, and have no restrictions. I have been surprised on a number of occasions to learn that a mature worker has never learned to drive. Many who have lived all their lives in large cities have never had any need to own a car or have never been able to afford to own one. Some men who can operate other machines very skillfully are in this category.

Don't overload passengers cars; they are not designed to substitute for trucks. I well remember agreeing to take a "few" items to another city, only to find that the car was totally unmanageable at turnpike speeds. In order to proceed or return, I had to repack the load, putting enough weight on the seats to make it possible to steer. Just because a trunk is big enough to carry a few hundred extra pounds does not mean it is safe to load it up.

Workers often expect company vehicles to be in bad condition. They think they are expected to put up with poor brakes, bald tires, worn shocks, and the like. They assume their foreman already knows about such defects, whereas he may be totally unaware of them. Vehicle safety should be an integral part of the safety program.

6.2.2 Crane Accidents

Accidents can happen while cranes are being assembled, moved, or prepared for moving to the next location, or simply parked, but I have chosen to deal with cranes as part of the materials-handling operations because the primary function of cranes is to move materials or subassemblies.

Crane operators are often required to be licensed, and each jurisdiction's regulations are likely to be different. Some licenses allow the worker to operate any type or size crane, derrick, or hoisting equipment. Elsewhere, the license may be applicable only to a designated piece of equipment. New York State has a law that applies everywhere in the state except in cities of over 1,000,000 population, provided such cities choose to write their own rules. New York City has done so, and has one license for cherry picker operators, one for crane operators when the boom is over 200 feet long, and a general license for all other operators.

Most licensing laws require that candidates be examined to be sure they do not have certain physical handicaps or illnesses such as epilepsy, heart trouble, hearing impairment, or uncorrectable vision. Wearing of bifocal glasses may be prohibited.

I agree with the intent of such rules; nobody should operate a crane if he is not physically fit to do the job. But there are obvious weaknesses in the licensing rules because a person's condition can change at any time.

Making it mandatory that an operator take a physical examination before being hired might be discriminatory unless all workers were required to pass an examination at prescribed intervals and it was a requirement that they be examined before being hired. In view of the importance of having people who are able to do their jobs without getting hurt or injuring others, I think the screening examinations are appropriate. They are very commonly required in other industries and the only problem I see in applying the requirement to the construction industry is the fragmentation of the work force into employees of many contractors and subcontractors, some of which may be on the job for only a short time. Difficulties of this nature should not, however, be an insuperable obstacle to overcome in order to be sure that crane operators are well enough to do their jobs.

But even preemployment screening examinations cannot guarantee that an operator is capable of working on any given day. Everybody has "off" days. It behooves supervisors to be observant, and note whether an operating engineer is on the ball. If an accident or a near miss occurs, I would recommend that the operator be sent to a doctor immediately for his own protection and the safety of everyone who might be affected by an accident to his crane. It may be that he has an illness that produces temporary dizziness, affects his coordination, or his ability to judge distance. He may be taking prescription drugs and getting an allergic reaction to them. Nobody wants to penalize a worker for conditions beyond his control, but it is not safe to have anyone operating a crane who is not up to the requirements of the job.

Of course there is the other possibility that a licensed operator may be taking illegal drugs, be an alcoholic, or have concealed a physical condition that would have disqualified him from obtaining a license. If such is the case, it is certainly appropriate to send him to a doctor whenever his performance is not up to standard.

Note that I recommend sending the worker to a doctor because his performance is substandard, and not because he is being accused of any wrongdoing. Even if he appears to be dead drunk, there should be no accusation of that nature; let a doctor make that diagnosis and be prepared to support his findings if called upon to do so in a labor dispute if the man is discharged.

Crane ratings include a safety factor, but the same factor of safety cannot be used to cover an overload, excessive reach, and a soft soil condition under the treads, all at the same time. Membership in a union of operating engineers does not make a crane operator a professional engineer who is qualified to revise the rating of a crane to compensate for unusual job conditions. If a special job or any other change in the configuration of a crane is made, it should not be left to the judgment of the operator to modify the rating. An operator may be reluctant to ask for help in determining the maximum load at different reach distances; that assistance should be given before he needs it.

Providing extra help to a skilled operator may take tact; suggesting that an extra man be assigned to watch for obstructions during a difficult move may be taken as lack of confidence in the worker's judgment. It is only natural to resent being warned against obvious obstructions and constantly told to be careful. But if an extra pair of eyes are needed to allow the operator to concentrate on the business at hand, any foreman worth his salt can assign the helper and give the operator's self-esteem a boost at the same time.

An extra signalman is often needed when a crane must pass under a power line. While traveling, the operator may have to watch for soft ground, open trenches, traffic, or absentminded pedestrians, and still keep his eyes open for overhead wires. It is very difficult to estimate the distance to a single horizontal wire unless there is some point of reference. A person looking backward past a wire at a moving crane has a better chance to judge the distance than the operator who sees only the wire against the open sky.

It is possible for a crane to strike an arc without actually touching a live wire; the arc can jump several feet if the boom approaches a high voltage line. Moreover, if it is going to happen, there is no time for evasive action. Electricity moves fast, as everyone knows.

When a crane must be moved, try to choose a route that avoids the need to pass under power lines if possible even if a longer distance must be traversed. If this is not possible, try to have the power turned off during the move, and to have someone from the power company present to confirm that the line is actually dead. If both these alternatives are not feasible, consider erecting guard ropes over the route of the crane a safe distance below and before the power lines, so that if the boom is too high for safety it will strike the guard ropes before approaching within arcing distance of the wires.

Work near power lines so often results in accidents that safety engineers and foremen should be completely familiar with the requirements of 29CFR 1926.550(a) (15), in which the subparagraphs dealing with clearances read:

> (i) For lines rated 50 kV. or below, minimum clearance between the lines and any part of the crane or load shall be 10 feet;
> (ii) For lines rated over 50 kV., minimum clearance between the lines and any part of the crane or load shall be 10 feet plus 0.4 inch for each 1 kV. over 50 kV., or twice the length of the line insulator, but never less than 10 feet;
> (iii) In transit with no load and boom lowered, the equipment clearance shall be a minimum of 4 feet for voltages less than 50 kV., and 10 feet for voltages over 50 kV. up to and including 345 kV., and 16 feet for voltages up to and including 750 kV.

Always consider every overhead wire to be energized until the person or utility owning it has visibly grounded it and has assured you that the line is dead and will not be reenergized while the crane is working in the vicinity.

Not long ago the operator of a cherry picker was electrocuted when his machine struck a power line. An employee of a different contractor tried to go to his aid, and was severely injured. He sued the owner of the project and the crane operator's employer for negligence and the court awarded him $375,000 damages. This set off a chain of counterclaims, and the litigation is still going on at the time I am writing.

If the boom of a crane passes near a powerful radio or TV broadcasting antenna, it can pick up a high radio frequency voltage. This can occur at a much longer distance from the antenna than would be dangerous at ordinary power line frequency. The engineer in charge of operations at the radio station should be consulted regarding this possible safety hazard. He may be able to suspend broadcasting for a few moments, or the move may be scheduled during time when the station is normally off the air. Regardless of the solution chosen, the importance of recognizing the hazard should not be overlooked.

The foregoing comments have been based on the assumption that qualified crane operators have been hired and that they abide by commonsense safety rules. They do not double block lines, swing loads over the heads of other workers, or leave the controls while the engine is running and before all brakes are set. They use their machines within their load limits, and inspect them carefully before starting work. They maintain good visibility and are insistent on having a solid footing in order to assure stability. They foresee possible accidents and take steps to prevent them. (Figure 6.1) Licensed operators can be expected to know their jobs; their errors are seldom the result of ignorance.

But, to be realistic, not all people who operate cranes have been properly

Figure 6.1 What if . . . ? To avoid accidents it helps if safety engineers and machine operators habitually ask themselves "What if . . . ?." Here, cranes were parked on a sunny Saturday with their booms and a suspended ball directly over the boardwalk at Atlantic City. Nothing happened; nobody was injured. But what if some prankster had disturbed the cranes?

instructed; they may never have had safety rules explained to them. The Construction Industry Manufacturers Association, 111 East Wisconsin Avenue, Milwaukee, Wisconsin 53202 has published some excellent booklets on basic crane safety. They would be suitable for distribution at safety meetings.

The American National Standards Institute (ANSI) has published Hand Signal Standards (ANSI B 30.5) that are nationally recognized and are approved by OSHA. If an operator and a signalman always worked together and could be certain of doing so for the rest of their lives, it would make no difference what signals they chose to use, provided they understood each other, but operators must expect to move from job to job, and it is therefore important that standard signals be consistently used. The standard emergency stop signal is made by extending one arm and moving the palm of the hand rapidly from right to left. It should be heeded by crane operators, regardless of who gives the signal, but with this sole exception, no other signal should be given an operator by anyone other than his assigned signalman.

Some comments were made in Chapter 2 about maintenance of excavating machines; they apply also to cranes. Be sure maintenance workers have the necessary manuals, tools, and access to spare parts. Above all, be sure they know how to proceed. They can ruin a new wire rope while unrolling and reeving it, they may make unsafe end connections, or may be injured if they fail to block or otherwise support parts on which they are working.

OSHA regulations regarding cranes, derricks, hoists, elevators, and conveyors are in 29CFR 1926, subpart N, beginning with paragraph 1926.550 and the latest revision of those rules should be reviewed by safety engineers before crane operations begin. Some highlights of regulations in force at this time are:

1. ANSI hand signals must be posted on the site and used.
2. Equipment must be inspected and repaired, if necessary, before each use.
3. Hoisting equipment must be inspected annually by a "competent person or by a government or private agency recognized by the U.S. Department of Labor." Records of such inspections must be kept.
4. Criteria for replacement of wire ropes are given.
5. It is mandatory that the accessible areas within the swing radius of the rotating superstructure of cranes be barricaded.
6. Every cab must have a 5BC or larger fire extinguisher in an accessible location.
7. There are rules for working near power lines and radio transmission towers.
8. Mandatory construction features of equipment are described, and many more are incorporated by reference to industry standards.

OSHA regulations governing the use of chain, wire rope, and fiber rope, and fiber rope slings are very specific; they are in 29CFR 1926.251 and 29CFR 1910.184, the latter being part of the general industry regulations identified as applicable to construction. A few highlights may be of interest, but the complete regulations should be understood by safety engineers.

1. Inspection of rigging equipment prior to use on each shift and as necessary during use is required.
2. Job fabricated hooks and makeshift fastenings are not permitted.
3. Chain must not be used if the wear at any point exceeds the amount indicated in table H-2 of 29CFR 1926.251. The wear permitted is of the order of magnitude of 20% of the original metal thickness.
4. Wire rope shall not be used if in any length of eight diameters the total number of visible broken wires exceeds 10% of the total number of wires in the rope, or it shows other signs of excessive wear, corrosion, or damage. In running ropes of cranes or derricks, wire rope must be taken out of service if there are six or more randomly distrubuted broken wires in one lay or three broken wires in one strand in one lay. See 29CFR 1926.550(a) (7) for additional rules applicable to cranes and derricks.
5. Knots may not be used to make eyes in wire rope bridles or slings.
6. Natural and synthetic fiber rope slings must be removed from service if there is abnormal wear, powdered fiber between strands, broken or cut strands, variations of size or roundness of strands, discoloration, rotting, or distortion of the sling hardware.
7. Wire ropes used for hoisting or pulling loads must be of one piece without knots or splices. When clips are permitted, the U bolt must go around the dead end of the rope; the number of clips that must be used is given in paragraph 1926.251(c) (5) table H 20, as follows:

Number and Spacing of U-Bolt Wire Rope Clips

Improved Plow Steel, Rope Diameter Inches	Number of Clips		Minimum Spacing (Inches)
	Drop Forged	Other Material	
½	3	4	3
⅝	3	4	3¾
¾	4	5	4½
⅞	4	5	5¼
1	5	6	6
1⅛	6	6	6¾
1¼	6	7	7½
1⅜	7	7	8¼
1½	7	8	9

8. Wire rope slings may not be shortened with knots, bolts, or other makeshift devices. They may not be pulled from under a load that is resting on them, and they must be protected from sharp edges of the load.
9. Temperatures above and below which various types of slings may be used are regulated.
10. Load limits for the commonly used types of slings and for several angles between the sling and the vertical are tabulated. See 29CFR 1926.251, tables H1 through H18.

Accidents have happened because workers failed to recognize that a load that was secure during transit may be unstable while it is being unloaded. The tailgate and part of the bed of a truck overhang the rear wheels, and if heavy materials are removed from the front of the bed before other heavy loads are lifted from the rear, the front of the truck may be lifted off the ground.

Now that the metric system is being more widely used, accidents can be expected to result from errors of arithmetic when converting from kilograms to pounds or tons. A worker may know that a crane and sling can lift 4000 pounds, but if a pallet has 20 bags on it, each weighing 50 kilos, does the load weigh half a ton, about one ton, or over two tons? In this case, since 1 kilogram weights 2.2 pounds, the lift would weigh 2200 pounds and could be safely lifted by a 4000 pound capacity crane and sling assembly, but the possibilities of error are apparent.

When the wind is strong or gusty, and the operator is finding it difficult to work safely, he is best qualified to decide whether work should be suspended. However, he may be hesitant to make the suggestion and his foreman should be alert to the need to discuss the situation with him. Two men died when a tower crane fell into the Columbia River at the Glen Jackson Bridge project because a support beam failed during a blustery wind storm.

If a crane has a jib, wind can cause the load to swing and exert a twisting force on the boom that is sufficient to cause failure.

I am sure that many people who are interested in crane safety are disappointed by the failure of accident reports to indicate why a failure occurred. It is often reported that a boom broke, but the reason is seldom given. Was it overloaded, had it been damaged in a previous accident, or had it deteriorated as a result of simple neglect?

6.2.3 Manual Lifting and Carrying

In spite of today's high labor rates, workers may still be seen carrying materials up ladders or hauling them up with ropes, a few pounds at a time. Both operations are costly, but carrying anything up a ladder should be prohibited; it is a common cause of falls and injuries to other workers. Anything that can be carried up a ladder can be put in a bag or bucket and hauled up with a rope, or better still, be moved by mechanical equipment. Roller and belt conveyors are so common in manufacturing plants and even supermarkets that people take them for granted, but few contractors have recognized that they could often cut costs as well as reduce the possibility of accidents.

Whenever workers are hired for jobs that require heavy lifting, they should be checked by a doctor. If a worker is hired who has a hernia or has back problems, a worker's compensation claim may have to be paid after he has been on the job a very short time. Workers who have not done much lifting should be trained to use proven methods. The National Safety Council, 444 North Michigan Avenue, Chicago, IL 60611 has published several booklets and films about lifting and handling methods. They stress proper positioning of the hands and feet, keeping the back straight, and using the strong muscles of the legs most effectively.

Incidentally, I have been told by doctors that they have no sure way to tell whether a person actually has back pain if there is no apparent injury. It is not too unusual for a worker to claim a back injury a few days before he expects to be laid off, and in some cases the same worker has been caught playing ball by an insurance investigator. Nobody wants to deny a worker financial benefits to which he is clearly entitled but, on the other hand, nobody wants to be taken in by a fraudulent claim. Get a detailed story of just what the worker was doing when the injury occurred, get witnesses to confirm it, and check the story carefully for inconsistencies. The actual injury may have occurred off the job at home or in an auto accident.

Hand injuries often occur while workers are handling pipe. Burrs inside the pipe remain if a rotary cutter has been used, and sharp threads can also cause cuts. Fingers can get caught between lengths of pipe. I wonder why pipe sticks are not more widely used. They can be easily made from metal or hardwood rounds with a collar set about six inches from one end. Workers standing at opposite ends of a pile of pipe can lift off individual lengths without danger of injury and with minimum effort (see Figure 6.2).

Hand injuries can occur when a worker passes too close to an obstruction or wall while using a wheelbarrow. Handles should have knuckle guards; they are easily made from scraps of 2×4, as shown in Figure 6.3.

The possibility of a tumble does not seem to deter workers from running a wheelbarrow on an insecure plank or from using one that has a cracked handle. Inexperienced laborers can be seen walking along with the handles of an empty wheelbarrow at eye level, so that if the wheel hits an obstruction they are almost certain to get a painful face, neck, or chest injury. I suppose that it is natural for foremen to pay most attention to workers who are

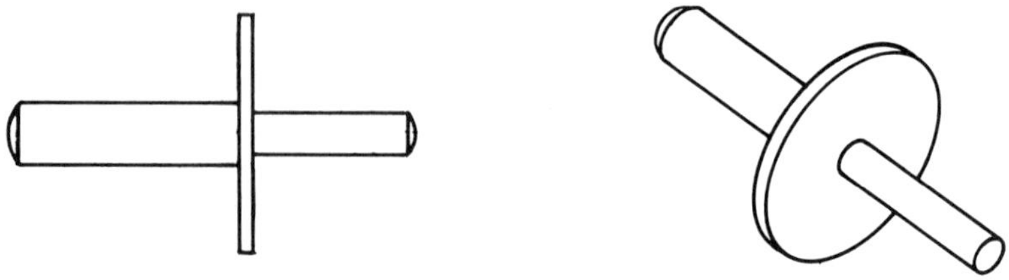

Figure 6.2 Pipe stick. This simple tool, easily made from hard wood or metal, is useful in avoiding hand injuries when pipe must be stacked or lifted from a pile.

using mechanical equipment, but the laborer who does not keep his mind on what he is doing can also get hurt on the job.

Thoughtless acts can cause accidents. A few days ago I noticed a worker toss a short piece of 2×4 to another who was standing on a beam about 30 feet above the ground. His aim was poor, the receiver reached for it, missed, but managed to keep his balance. The piece of wood narrowly missed two other workers on the next lower level. I did not hear their comments but, surprisingly, just a moment later, the same worker tossed another identical piece of wood to the same receiver, and everybody seemed to take it in their stride. It was accepted as a normal procedure. Soon thereafter, a man with a plumb bob asked a worker on the next lower level to toss something up to him between the bar joists on which he was standing. Again, poor aim resulted in the object hitting one of the joists and nearly hitting the one who had tossed it as it fell. They tried again, with the same result, and finally made it on the third try. Neither thought of the possibility of tying it to the cord of the plumb bob that was dangling between them.

About 6% of construction fatalities are caused by the worker being struck by some object.

Metal pails with wire bail handles are often reused to handle small items such as nuts, bolts, or tools, without considering the weight to be lifted. Handles are apt to have been damaged in transit, and at best are not likely to be much stronger than absolutely necessary to support the original

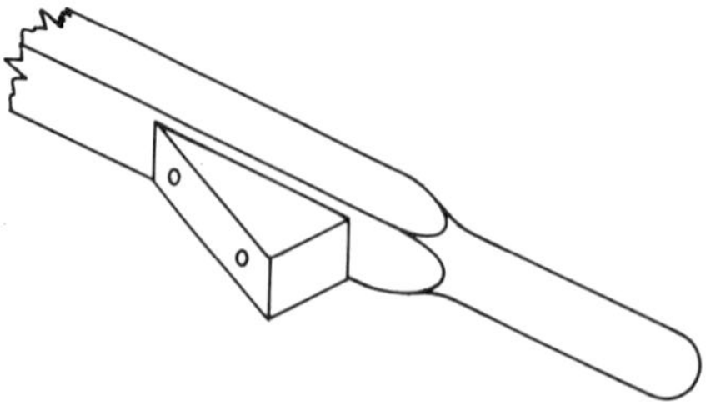

Figure 6.3 Knuckle guard for wheel barrow. A wooden block on each handle will avoid injury if the barrow swerves against some obstruction that could crush the worker's hand.

contents of the pail. Even less secure are the flat metal handles that are soldered to the tops of rectangular metal cans. They are not to be trusted even when new if the can is the five gallon size. Sooner or later, one of the cans will be received with a handle that is not securely attached and, when the poor joint fails, the sharp end of the handle is sure to inflict a painful cut on the worker's hand. Still, I have often seen cans of this type reused, with half the top cut away, to carry welding rods or metal parts, often far heavier than the liquid the can originally contained.

I think far too little attention is given to the selection and use of small diameter rope used to lift materials and tools by hand. Many accidents occur when men carry things in their hands while climbing a ladder; they should use a rope to bring up whatever they need and give their full attention to the ladder while they are climbing.

Rope used for lifting objects or as a life line should be selected with care, used and cared for properly, and cut up when it is no longer safe to use so that some unsuspecting person will not hurt himself or others while relying upon it. Many different types of rope are available, and it is worth the trouble to choose the right one.

Manilla, the old favorite, is likely to remain popular. It has a rough enough surface to make it easy to hold, and knots do not tend to slip excessively. It does, however, absorb moisture readily, and this can make it difficult to untie knots that were tied in the dry line. If the rope is not dried out before being laid aside, the fibers may rot and lose practically all their strength.

Sisal, a cheaper substitute for manilla, has less initial strength, is very subject to deterioration, and is harsh or splintery to handle.

For construction work, the synthetic fiber ropes are probably the best choice, if properly selected with due regard to the peculiarities of each material.

Nylon is often an excellent choice. Size for size, it is over twice as strong as manilla, is highly resistant to deterioration by moisture, and is noted for its elasticity. When highly stressed, it will stretch before breaking. On the other hand, the surface friction is lower than manilla, making it necessary to select the right knot carefully to be sure the free end will not pull through the knot. Also, if an overloaded rope does break, the stored energy that is released makes the free ends whip about with dangerous force.

Polyester fibers, of which Dacron is most often used for rope manufacture, is about as strong as nylon, but it has considerably less tendency to stretch under load. This property may be desirable when there may be changes in the rope tension and exact positioning is important, or when there is a possibility the rope may be broken and reducing the danger of injury by the loose ends is desirable. It is rot resistant and is not greatly affected by water.

Polypropylene is stronger than manilla, but considerably weaker than either nylon or Dacron. It has two serious deficiencies. The ultraviolet rays in sunlight cause it to deteriorate, and to minimize this effect the naturally colorless material is generally mixed with yellow, black, or other pigments. The second drawback to its use is the low surface friction; knots tend to come untied very readily. A steeplejack fell to his death when the short end of the rope supporting his bosun's chair slipped through the knot, and several other fatalities have been caused by similar occurrences. Free ends of polypropylene should have a stopper knot to prevent them from passing through another knot, or the free end should be served or taped to the stressed part of the line. Similar precautions are advisable with all the synthetic fiber ropes, but polypropylene is particularly prone to slippage.

Polyethylene is used to make small rope, generally about 5/16 inch in diameter, and it is usually brightly colored. It has the poor surface friction of polypropylene, and is readily weakened by sunlight. It is useful for barricades, but is not as suitable for the manufacture of rope as the other materials I have mentioned. It does have one feature that may be useful; it will float.

When the synthetic ropes touch hot welds or are heated above about 180°F, their strength is reduced. Melting of the fibers will permanently damage the rope, and any spots where damage has occurred should be cut out and discarded.

Watch the strands of polypropylene and polyethylene ropes carefully for broken fibers; they are sharp and can cause painful abrasions. In this respect, nylon and Dacron are preferable.

Safety engineers should not assume workers know how to tie even the simplest knots. It is better to assume ignorance than competence; a worker may resent being asked to demonstrate that he knows how to attach a rope to a load, but if he does it in an insecure manner, there may be a serious accident. People who have been accustomed to using only the natural fiber ropes may make dangerous mistakes when they begin to use the synthetics. Valuable time is also lost when a knot is chosen that is difficult to untie. A good knot should be secure, but easily loosened. Many hundreds of knots have been devised for specific purposes, and I recommend that workers be required to use certain ones for specific purposes rather than being permitted to make their own choice or to "invent" a new knot when they consider it necessary. Often an improvised knot behaves in a manner that the user did not anticipate. Even the well-known knots may have surprising characteristics.

For example, practically everybody knows how to tie a square or reef knot, although the granny knot that so closely resembles it still shows up occasionally. A dangerous popular belief is that a square knot can be used for just about every application and in his classic 600 page book on knots, Clifford W. Ashley says that when used to tie two ropes together, the square knot has probably caused more deaths and injuries than all other knots combined.

Never use a square knot to join two ropes of different sizes or materials;

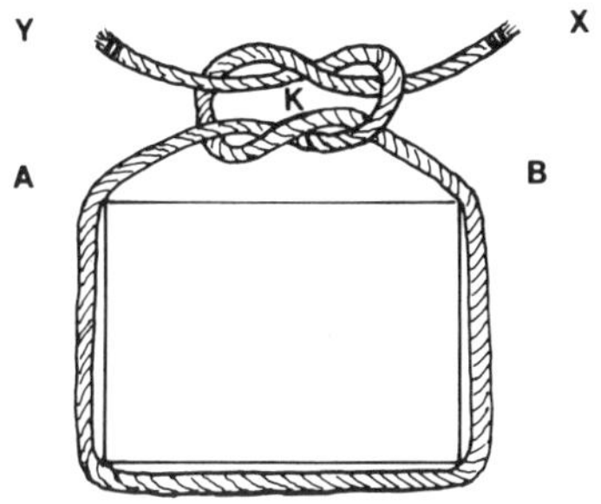

Figure 6.4 Square knot. Although it is popularly considered reliable, the square knot is very apt to fail if improperly used. It is not suitable for joining ropes of different sizes, materials, or stiffness. Be sure the knot is properly formed and tightened, with KB and KX side by side as they pass under the righthand loop. Never lift a load or slide it by pulling on X or Y; the knot may be readily untied in this manner unless it has been pulled up very tightly.

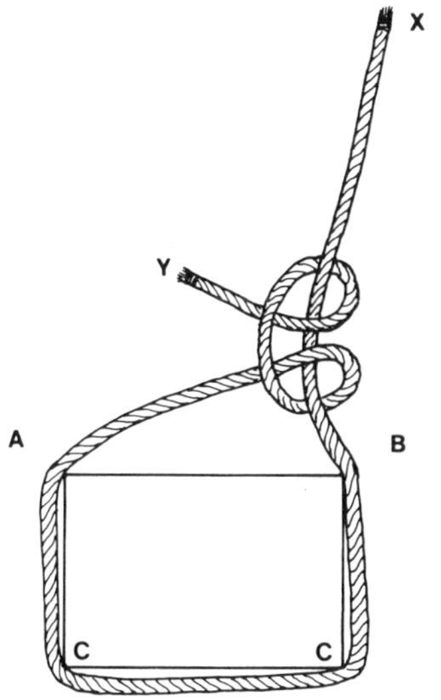

Figure 6.5 Upset square knot: **DANGER.** Pulling on end X has upset the square knot shown in Figure 6.4 and that part of the rope between B and X can slip freely through the two loops ending at Y. The load will fall if an attempt is made to lift it by placing a hook between A and B under the knot. Always protect rope from sharp corners, as at C.

it is very apt to come undone or to slip under these conditions. Even when the two ropes are identical, the squre knot is a poor choice, particularly when the synthetic fiber materials are used. It was used on sailing ships to tie reef points because it could be very quickly untied by jerking one of the free ends when sails were to be unfurled. This feature can be readily demonstrated at a tool box meeting. Tie a piece of small diameter rope around a box, as in Figure 6.4. Slip one hand between the rope and the box and note how the knot tightens when strained. If synthetic fiber rope is used, its tendency to slip through the knot can be shown. Short ends X or Y may slide through the square knot before it pulls up tight.

Now, having tied a normal square knot, pull on either of the free ends and watch it upset to the configuration shown in Figure 6.5, which in the illustration is the result of pulling on end X of Figure 6.4. Now put one hand under the rope and try to lift the box; the good old square knot is no longer there. If lifting the load by pulling on end Y is attempted, end X will simply slide through the two loops. If the box is lifted by pulling on end X, the knot will slide along the rope, tipping the load and squeezing it.

Figure 6.6 Bowline knot. This is one of the most reliable knots, and is often used as a substitute for an eye splice. When so used, it can develop about 50% of the rope strength, whereas an eye splice over a thimble can have as much as 90% of the strength of the rope. Be sure to leave an adequately long tail at E to allow for slippage. Two ropes of different characteristics may be joined by tying bowline knots at their ends, with the loops interlocked.

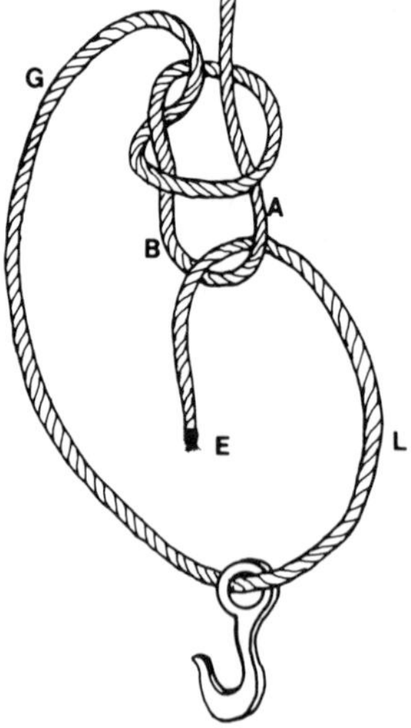

Figure 6.7 Upset bowline: **DANGER.** In Figure 6.6 if a strain is put on the section of the knot between E and L while S is slack, a loosely tied bowline may be deformed to the *unsafe* configuration shown above. End E can slip through loop A-B and drop the load.

The upset bowline consists of a slip knot with tail E through the loop. If a strain is taken on end S, the upset bowline can be converted to the standard form; this provides a quick method of tying the bowline. With practice, it can be done with surprising speed.

The bowline knot is easy to tie, easy to untie, and when properly used it is very secure. It is the preferred substitute for an eye splice and ropes of different sizes can be safely joined by a bowline knot in the end of each. For maximum security when using a rope that has a low friction surface, tie a stopper knot in the short free end, or serve it to the adjacent part of the line. Figure 6.6 shows the stable form of the knot; it is secure when the strain is on the end marked S.

To demonstrate how the bowline can be upset to a dangerous configuration, pull on ends E and L, thereby converting the knot to the configuration shown in Figure 6.7. It now consists of a slip knot with the part of the rope marked E-L through the loop. Disengage that part of the line, pull on G and S at the same time, and in a flash the knot is untied. Always check knots to be sure they are in the proper form before relying upon them.

The larks head stoppered, Figure 6.8 provides a quick and secure method of attaching a line to a pail or tool; it is used in a manner similar to a choker sling. The long loop, L, can be formed by splicing, or a bowline may be used. The overhand knot shown is also suitable, but is difficult to untie. I show it because even unskilled workers instinctively know how to tie it.

To attach a bucket, slip the long loop under the bail, spread the loop, and put the whole bucket through it. Pulling on free end S tightens the connection. Note that it is not necessary to disturb the knot or splice that forms the long loop when attaching or detaching the pail. Watch for con-

Figure 6.8 Larks head, stoppered. This knot, used as illustrated, tightens around the load like a choker sling. Knot A need not be united to attach a pail to loop L provided it is long enough to slip around the load. See text for further comments.

centrated wear at the end of the loop and cut off the defective part before it seriously weakens the loop.

Two half hitches can be tied as in Figure 6.9. I have not found it makes any difference which way the second hitch is tied relative to the first. The looped modification is also commonly used. Inexperienced workers tend to add more and more hitches to make doubly sure the knot will hold, but two are enough, provided the free end E is long enough to permit normal slippage to occur as the knot tightens. For greater security, tie a stopper knot at the free end, or serve it to the strained rope.

Tag lines should be used whenever a load is lifted. If this is standard procedure, it will be easy to control the load if it begins to swing or is about to hit an obstruction.

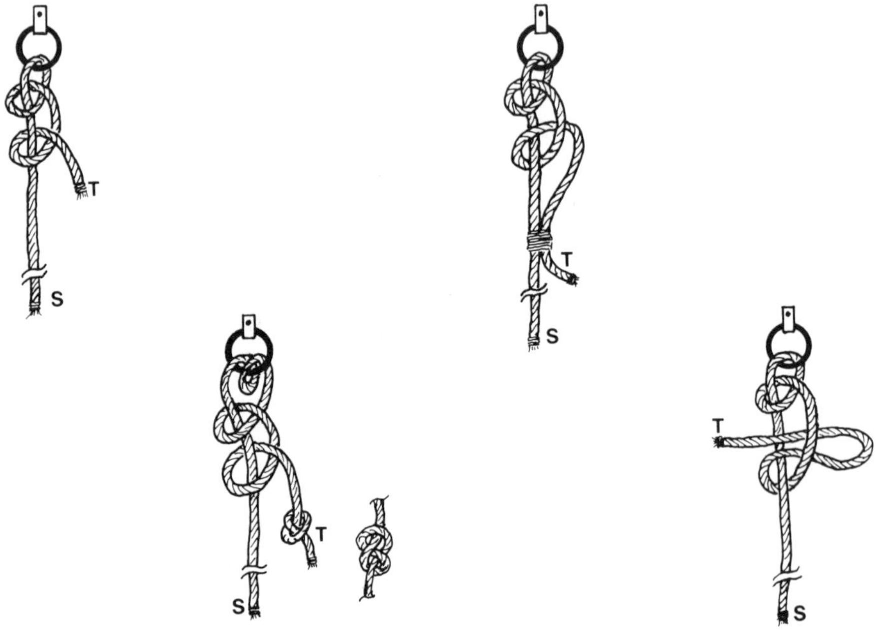

Figure 6.9 Two half hitches. Several variations of two half hitches are possible, and they are known by many different names. The common form, upper left, tightens around the ring or load while there is a strain on S, but repeated shaking or alternate slacking and jerking may cause the knot to come untied, particularly if one of the synthetic fiber ropes is used. Be sure to leave a long enough tail, T, to permit some slippage as the knot tightens under load.

The illustration at the upper right shows the second loop reversed, which has little effect on the characteristics of the basic knot. The tail has been served or taped to S to prevent the knot from coming untied.

The lower left sketch shows a double turn around the load or through the ring, which makes the knot somewhat easier to untie. A simple overhand stopper knot has been tied in the tail to keep it from slipping through the half hitches. An alternative stopper knot, the figure eight knot, is also shown.

For ease of untying, the knot shown at the lower right is handy, but of course it is unsuitable if the strain is variable and could shake the knot loose.

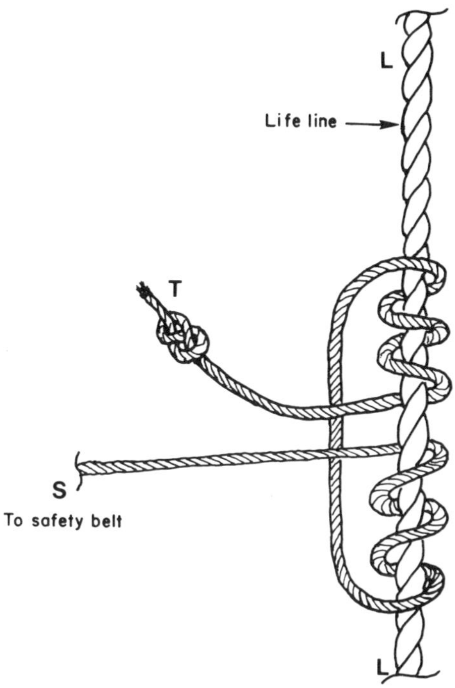

Figure 6.10 The triple sliding hitch is useful for attaching a safety belt lanyard to a life line, L-L. Note the upper and lower turns are in opposite directions but both ends, S and T are on the same side of the vertical part of the knot. If loosened by twisting the two sections, the knot can slide easily along the life line but it will strongly resist movement when tightened and a strain comes on either free end. Be sure end T is over a foot long and keep the distance to the safety belt as short as possible.

The ends of all ropes, regardless of size, should be siezed, back spliced, or otherwise protected from fraying. Some synthetic ropes can be fused at their ends by heat; check with the supplier whether this is recommended and how best to do it.

There are many ways to attach a lanyard to a life line, and suppliers of safety equipment offer many devices. If it is necessary to use a knot, the one illustrated in Figure 6.10 has the advantage of security and ease of adjustment. By twisting the upper and lower turns of the knot, it can be loosened enough to make it easy to slide along the life line without having to untie it.

Note. *To prevent slipping, be sure to tighten all knots before putting strain on them. When using synthetic fiber ropes, leave an extra long tail in anticipation of some slippage. If the strain on the knot will be jerky or if slippage is apparent, tie a stopper knot in the tail or serve it to the adjacent strained part of the line.*

6.2.4 *Placing Concrete*

Some comments have already been made about placing concrete footings and foundation walls in Chapter 5, Section 5.10. Good construction practices and safe operations continue to supplement each other while the rest of the concrete is being placed. The stress on forms is, of course, increased while vibrators are being used, and poor design or construction is apt to cause failures at such times. Workers not actively participating should be kept clear of the operation.

Before forms are to be stripped or moved, safety precautions should be reviewed with workers. Accidents at elevated structures are always a possibility.

In 1978 a large part of a cooling tower form collapsed at a power plant in Willow Island, West Virginia, killing 51 workers.

In March, 1981, two men were killed, a third man was seriously injured, and a fourth narrowly escaped injury when rods failed to secure a 30 × 6 foot steel form section of a cooling tower. One rod was reported to have been improperly secured and, when it came loose, three others broke as a result of excessive loading.

Early in 1981, in Vancouver, British Columbia, a flying form weighing over three tons fell 36 stories to the street, carrying four workers to their deaths.

In 1978, during a tornado, a crane collapsed inside a tower being constructed for the Mississippi Power and Light Company. It knocked out a section of concrete 70 × 150 feet. The damaged parts were cut out and trimmed with diamond saws and replaced.

These accidents did not happen to fly-by-night contractors; all were experienced. But it only takes one error of judgment, design defect, or weak spot in the structure to cause a tragic accident.

Precast concrete failures may occur either during or after the sections have been put in place. A typical failure of a T beam caused a 27 × 60 foot section of a parking garage to collapse, smashing 18 cars but causing no injuries. The beam that initiated the collapse was found to have been cracked.

Defects of this nature may be caused by improper design that failed to anticipate stresses that would be imposed during erection as well as after the structure was completed. Omission or improper placement of steel in the form or incorrect tensioning may cause failure. Rough handling en route to the job can also weaken precast assemblies. The primary responsibility for failures of this nature lies with the project manager, but the safety engineer has an interest in it also. One must keep an accident from happening, the other must see to it that if a collapse occurs, nobody gets hurt.

6.3 Falls

Employer's Insurance of Wausau, Wausau, Wisconsin has made detailed injury cost analyses, based on payments made to injured workers in various worker's compensation classifications. Table 6.1 is based on their studies. For the job classifications shown, from 26 to 46% of compensation payments were made to workers who were injured in falls. Clearly, falls are an important problem to anyone interested in reducing the cost of accidents. The incidents covered in the table included 16,873 cases, and the average payments for the different occupations varied from $1600 to $6029 per injury. These are direct cost figures. If indirect costs are included, based on the findings at the Department of Civil Engineering at Stanford University, total costs per accident would be from 4 to 17 times the direct costs.

I found it interesting that falls accounted for 46% of compensation payments to workers in classification 5059, iron or steel erection, frame structures not over two stories high. Obviously, falls are not confined to tall buildings. In fact, for classification 5057, iron and steel erection, N.O.C., the percentage of costs attributable to falls was slightly less, (41%), and the average cost per injury was higher ($6029 vs $2677), which indicates that fewer falls occurred from heights over two stories but there were more serious consequences when they did occur.

Not all falls occur from an elevation; many workers slip or trip over something on the working surface without falling from a height. Wausau identified these costs for a few classifications and reported that such falls accounted for 11% of the payments to electricians and 16% of the payments to plumbers.

TABLE 6.1. Percentage of Worker's Compensation Costs Due to Falls

Average Cost per Injury	Worker's Compensation Classification	Percent of Costs due to Falls
$6029	#5057—Iron and steel erection, N.O.C.	41%
$4267	#5213—Concrete construction, N.O.C.	27%
$3905	#5190—Electrical wiring	43%
$3231	#5551—Roofing	33%
$2677	#5059—Iron and steel erection, frame structure. Not over two stories	46%
$2323	#5183—Plumbing, N.O.C.	26%
$1600	#5538—Sheet metal erection, N.O.C.	34%

Source: Technical Guides, published by Employer's Insurance of Wausau, Wausau, Wisconsin, covering analysis of 16,873 injuries.

U.S. Department of Labor Statistics Report 586 includes records of fatal accidents to workers in the private sector. In one year, 925 construction workers were killed or one out of every 3300 included in the survey. Falls caused 29% of these fatalities, and no other industry had such a high incidence of falls.

Statistics tell only part of the story; accidents happen to people who are more than mere numbers. Let's look at a few typical case histories also recorded by Employer's Insurance of Wausau.

> A 42 year old mason set up a 10 foot high rolling scaffold but omitted the railing. He fell to the ground, injuring his spinal cord. The lower part of his body was paralyzed.
>
> A 40 year old man was overreaching when he lost his balance and fell 5½ feet from a stepladder. His head struck the base of a machine and he was pronounced dead on arrival at the hospital.
>
> A 20 year old painter fell 65 feet from a swing stage. He would have survived had he been wearing a safety belt with its lanyard attached to a life line.
>
> A carpenter, 61 years old, fell 14 feet through an unprotected opening in a floor. He suffered permanent paralysis from the waist down.
>
> An electrician, 27 years old, fell 20 feet and suffered multiple injuries including a fractured and dislocated hip and pelvis, ruptured spleen, and a dislocated elbow. He was coming down a ladder while carrying tools.
>
> A 25 year old painter failed to check his ladder to be sure it was secure. It slipped from under him, and he fell 16 feet onto a concrete slab, sustaining back and head injuries.

Note that these cases include men from 20 to 61 years old. No age group is immune to falls.

At a high rise office building an ironworker was killed by a fall from the fifth floor in July, 1980. In February, 1981, the contractor was given eight safety violation citations, but about a month later another ironworker fell from the sixteenth floor and was killed. He was wearing a safety belt, but had failed to attach the lanyard to anything. If accidents "will happen," as some people believe, why is it that some contractors lose more men than others?

There is room for disagreement about the best way to prevent falls, but with some reasonable measure of mutual respect on the part of the workers and the safety engineer a solution can generally be agreed upon. Unfortunately, people do not always want to work out their problems. In one instance, an OSHA compliance officer issued 18 citations at a job in Las Vegas. No accident had occurred, and only one condition was classified as "serious," defined as a hazard where there is a substantial probability that death or serious physical harm could result and the employer knew or should have known of the hazard. Workers were setting forms on the thirteenth floor level without protection against a potential fall of 130 feet. Instead of working out some reasonable method of protecting the workers, the matter became a confrontation. At a hearing before an administrative law judge appointed by the independent Occupational Safety and Health Administration Review Commission, the employer claimed that the compliance officer was hostile, inexperienced, and careless. A safety engineer employed by the firm that sold the forms to the contractor testified that use of life lines could have created some tripping hazards, but that in his opinion the danger of a 130 foot fall was a more serious hazard. The judge upheld the citation and the employer appealed again to the review commission. Once more he lost. In view of the relatively small $700 penalty, it seems that personal animosity may have obscured the main consideration, which should have been the safety of the workers.

OSHA regulations governing situations of the type involved in this case recognize that job conditions vary and they allow the contractor considerable leeway in selecting appropriate safety measures. The rule expressed in 29CFR 1926.105(a) reads:

> Safety nets shall be provided when workplaces are more than 25 feet above the ground or water surfaces, or other surfaces where the use of ladders, scaffolds, catch platforms, temporary floors, safety lines, or safety belts is impractical.

Safety nets do work; while the Golden Gate bridge in San Francisco was being built, 19 workers became members of the "Half Way to Hell Club" when they fell into the safety nets (Figure 6.11).

Other OSHA regulations affecting steel erection include the following highlights in paragraphs 1926.750 and 1926.751 of 29CFR.

> Permanent floors shall be installed so there is not more than eight stories between the erection floor and the uppermost permanent floor, except when structural integrity is maintained by the design.
>
> During skeleton steel erection, a tightly planked temporary floor shall be maintained within two stories or 30 feet, whichever is less, of that portion of each tier of beams on which any work is being performed.
>
> During skeleton steel erection, where the requirements of the preceding paragraph cannot be met, and where scaffolds are not used, safety nets shall be installed and maintained whenever the potential fall distance exceeds two stories or 25 feet.
>
> A safety railing of ½ inch wire rope or equivalent shall be installed around the perimeter of all temporarily floored buildings, approximately 42 inches high, during structural steel assembly.
>
> When placing structural members, the load line shall not be released until the member is secured by at least two bolts, or the equivalent, at each connection, drawn up wrench tight.

Figure 6.11 This safety life net was assembled from about 100 pieces of 1 inch high strength flat synthetic webbing each about 9x12 feet with 5x5 inch openings. The panels are laced together and to the structure to provide personnel protection and also access to connection points of a radar space tracking station. Individual panels can be readily removed for reuse. The assembly meets American National Standards Institute (ANSI) A 10.11 and OSHA 29CFR 1926.105 requirements. (Photo courtesy of Miller Equipment Company, Franklin, Pennsylvania 16323.)

But recognizing hazards and understanding regulations will not reduce accidents unless workers accept more personal responsibility for their own safety. Traffic must stop for school buses because toddlers have to be supervised at all times, but adults should not expect others to be totally responsible for their welfare. Mature construction workers should recognize situations when they need safety harnesses and should use them without having to be told to do so. If the work is such that they consider it unsafe to proceed and the employer refuses to take the appropriate steps to eliminate the danger, workers can always walk off the job until the matter is resolved one way or another.

Safety belts are one of the tools of the construction trade; workers should know about them, just as they should know the different kinds of wrenches and other tools. Four classes of belts or harnesses are listed by the American National Safety Institute in ANSI Standard A 10.14 1975.

Class I belts are used to restrain a person in a hazardous work position and to reduce the probability of falls. Figure 6.12 shows a rod man wearing a Class I assembly. Note the belt includes an optional broad pad for comfort, and the lanyard has snap hooks for convenience in attaching the chains to the rods. He has a good pair of gloves and his pliers are made with a hook bend on one handle to provide a better grip. He is well equipped to do his job. Class I belts are not intended to stop vertical falls; they are made to prevent them from happening, rather than to catch a worker in mid air.

Class II harnesses consist of a belt and chest straps. They are made to be used where there is a limited need to arrest a fall and are useful for pulling a person from a confined space, but in common with Class I belts, they are not intended to stop vertical free falls. Figure 6.13 shows a typical harness and its use by a welder.

Figure 6.12 This rod man is wearing an ANSI A 10.14-1975 Class I belt. Note that it has a wide pad across the back and lanyard connections that make it convenient for him to move about. The belt manufacturer is also well known as a source of tools such as pliers with a bent handle to minimize slippage or dropping the tool. Class I belts are intended to help the worker maintain a safe working position and thereby avoid falls rather than to arrest them. (Photo courtesy of Klein Tools, Inc., 7200 McCormick Blvd., Chicago, Illinois 60645.)

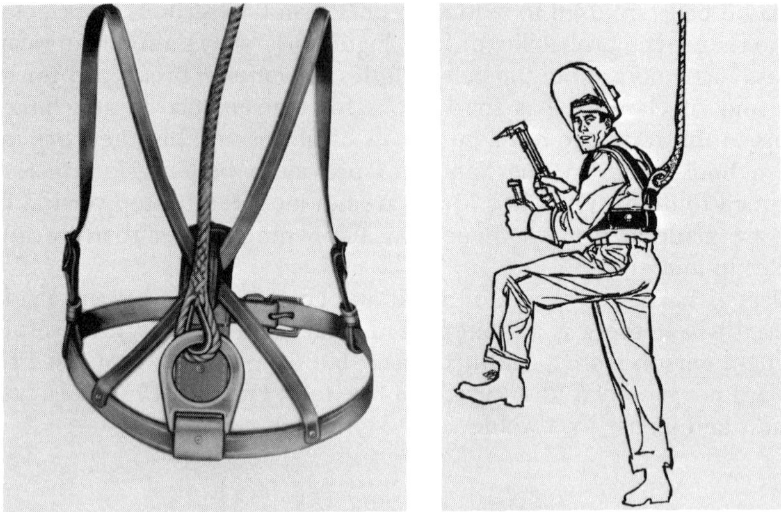

Figure 6.13 Class II harnesses consist of a belt and supplementary chest straps to provide limited capability to arrest falls. Note the lanyard attachment at welder's back has been made with an eye splice over a metal thimble (Illustration courtesy Klein Tools, Inc., Chicago, Illinois 60645.)

Class III body harnesses are designed to arrest severe free falls and can produce a stopping force of as much as 35 times the worker's weight. Figure 6.14 illustrates a typical arrangement to distribute the stopping force at the end of a fall over the thighs, chest, buttocks, and shoulders.

Class IV suspensions include bosun's chairs and harnesses for rescue workers. They are not fall-arresting devices; their purpose is to provide a safe support for workers. See Figure 6.15.

While OSHA requires employers to provide safety gear, I have heard of no law that prohibits a worker from buying it himself. I see no reason why an employer should deny him the privilege of wearing any approved model of appropriate class of safety harness he finds most comfortable. Similar privileges are common throughout industry. Salesmen often prefer to drive their own cars, and the mileage allowance given them by their employers may or may not cover the salesman's ownership expenses. It is expected that many professionals will have their own instruments, and machinists frequently own fine tools of which they are justifiably proud. It is perhaps a fair generalization to state that ownership of one's tools is a sign of professional status. If this is true, ownership of good safety gear could be accepted as indicative of a worker's qualifications to do hazardous and prestigious work.

That it is not now customary for construction workers to buy their own safety gear and to wear it proudly is of no particular significance. Customs

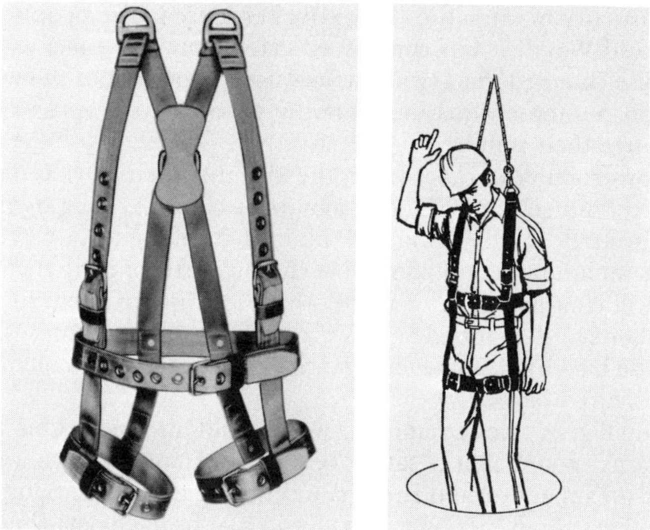

Figure 6.14 Class III body harnesses are designed to arrest severe falls and to enable a worker to be lifted from an enclosed space even though he may be unconscious. Note belt, chest, and thigh straps as well as double attachment to lanyard. (Illustrations courtesy Klein Tools, Inc., Chicago, Illinois 60645.)

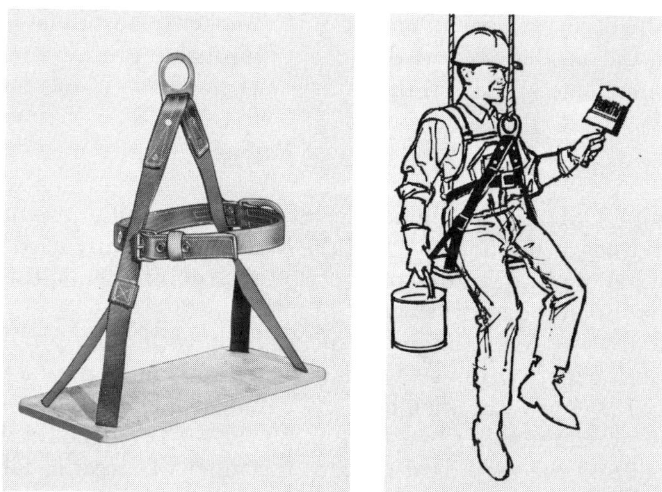

Figure 6.15 Class IV suspensions include the bosun's chair and similar equipment designed to provide a safe working support rather than to arrest falls. Note that the chair support is supplemented by a separate line attached to the worker's belt. (Illustrations courtesy Klein Tools, Inc., Chicago, Illinois 60645.)

change, particularly when the change makes sense to the people involved. Before World War I, it was considered effeminate to wear a wrist watch, but army life changed that. Hod carriers used to be part of the work force on every job, but those who thought nothing would ever replace them have had to change their minds.

Many construction workers foster the idea that their work is dangerous; there is a certain self-image they enjoy projecting. As long as the public accepts this image at face value and thinks that willingness to accept needless risks is an inherent and admirable characteristic of construction workers, they will continue to encourage others to think of them in that way. On the other hand, if the idea gets around that taking unnecessary risks is just a little bit stupid, some changes may occur. Nothing collapses faster than a punctured balloon.

Some intelligent sales promotion work could also help. One engineer, Ron Williams, noted that a safety belt he had borrowed to wear while putting up an amateur radio antenna was unsafe to use. He bought a good one, told others about his observations, and now he has a part-time business that retails more safety belts per year than many full-time distributors. He knows he is doing someone a favor when he sells a belt that could save the user's life.

I am confident that changes will occur and that wearing safety gear will one day be as macho as cowboy boots have been in recent years. But meanwhile safety engineers need to identify and correct hazards that could cause falls. It is an everyday job. I have listed some hazards in several parts of Appendix K, but many more will occur to anyone who uses his imagination as he observes work in progress. If a worker fell, would it be possible to state in the accident report that every reasonable precaution had been taken to anticipate and avoid the injury?

6.3.1 Guard Railings

Throughout OSHA regulations the terms "guard rails," "standard railings," "midrails," "toe boards," and the like are frequently used. In order to avoid citations for failure to meet specific requirements, their official definitions should be understood.

A standard wooden railing is 42 inches high, with the top rail made of a 2 × 4 and with 2 × 4 posts 8 feet on centers. It must have a toe board made of a 1 × 4 or wider member, and a midrail made of a 1 × 6, located halfway between the other two. Wire mesh (½″—#18) is required between the midrail and the toe board if there is danger of material falling onto anyone working below the scaffold. The posts and the top rail must be capable of resisting a 200 pound force in any direction, except that removable sections need not be anchored against an upward force. See 29CFR 1926.500(f).

Stair railings are similar to standard railings but the top rail must be smooth and be set 30 to 34 inches high, measured from the top of the rail to the forward edge of the stair tread. Exposed sides of stairs must have mid rails and meet the other requirements of standard railings.

Hand rails must clear other objects by about 3 inches.

The ends of all railings must not extend beyond posts in a manner that would create a projection hazard.

Steel railings may be made of 2 × 2 × ⅜ inch angles or of 1½ inch nominal size pipe. If job conditions make it necessary for a stronger railing to be provided, it must be installed.

Wall openings from which there is a drop of more than 4 feet require a railing, and open-sided floors or platforms must have one if there is a drop of 6 feet or more to the surrounding ground or floor. This applies to temporary floors as well as to permanent parts of the building, except that during steel erection 29CFR 1926.750 permits the use of only a single ½ inch wire rope around temporarily planked or temporarily steel decked floors.

Scaffolds over 10 feet high all require standard guard rails on all exposed sides and ends. They are also required on scaffolds 4–10 feet high if they are less than 45 inches wide or 45 inches long.

I recommend that safety engineers read the OSHA regulations mentioned previously; they include many details I have not included for lack of space. These rules are quite clear; there may occasionally be a question of interpretation but their intent is seldom obscure. Still, anyone who has read them cannot fail to be aware that they are commonly ignored until an accident occurs or a compliance officer issues a citation. Then there is apt to be disagreement on the need for the railings and quibbling about phrases in the rules.

I think OSHA should have a meeting of the minds with contractors and workers in order that railings that are needed may be voluntarily installed and that enforcement of regulations may be within the realm of possibility. It will never be possible as long as there is such apparent disagreement about what is necessary. Possibly a compromise may be reached whereby the upper rail will be considered sufficient in many instances. Alternatives to the standard railing might be mentioned in the regulations. I think that in many cases a single sheet of plywood or a piece of expanded metal with reenforced edges might be useful. There are probably many ways to keep workers from falling that would be acceptable to all concerned.

6.3.2 Scaffold Accidents

Injuries to workers who fall from scaffolds are very common, but most of the scaffold accidents that make newspaper headlines are caused by design or erection errors over which the workers who are killed or injured have no control. After an investigation that took five months, the California Division of Occupational Safety and Health found that design, construction, and supervisory shortcomings were all primary causes of failure of a complex scaffold and hoist assembly that failed at the Helms pumped storage power project in California. Seven workers were killed. Lack of training was a contributing cause of the accident; it was the first day on the job for two of the hoist operators and one of them could not maintain control of his hoist. Twelve citations were issued, six of which were "willful/ serious" ones, but the total fine was only $60,250.

A recent scaffold collapse in Chicago killed 12 workers. Insufficiently strong concrete pads on which it rested were blamed for the failure.

Accidents of this kind indicate that it is possible to underestimate the importance of sound design and construction of scaffolds. They are temporary structures, but they can get the equivalent of years of service in only a few weeks. They are exposed to weather and are often so heavily loaded that they may be stressed to their capacity with little or no safety factor. Repeated erection and dismantling also take their toll. Therefore, in contrast to the structural members of the finished building that are protected from the weather and seldom disturbed after erection, scaffolds get hard use.

Scaffolds should be erected and dismantled by experienced workers who know how to take care of themselves. The carpenter who builds a railing to protect others does not have comparable protection himself. 29CFR 1926.451(a)(3) requires that "no scaffold shall be erected, moved, dismantled, or altered except under the supervision of competent persons." That is a clear and comprehensive statement right at the beginning of the regulations about scaffolds. But following this good beginning, my copy of the rules has 20 pages of instructions that spell out to the "competent" person just about everything one might expect him to know already. This discourages the use of judgment in cases where heavier construction might be desirable. For example, the Cathedral of St. John the Divine in New York City has been under construction since 1892 and a "temporary" scaffold might be in use for many years on a project of this kind.

When the Prince of Wales and Princess Diane were married recently in St. Paul's in London, I recalled the story of a near accident to Sir John Thornhill while he was painting the pictures around the inside of the dome, 100 feet above the floor. He stepped back to admire his work, and was in imminent danger of falling off the scaffold when a quick-witted workman grabbed a brush and defaced the painting, an act he knew would be the only way to get the artist's attention in time to save his life. Incidentally, the scaffolding outside that edifice was in service for over 20 years and the poles were as thick as present day power poles. In fact, the building could scarcely be seen through the massive assembly of planks and framing.

I have combined nine tables from 29CFR 1926.451 into three equivalent summary tabulations in Appendix H. This should make the pertinent material easier to use and emphasize the differences between light, medium, and heavy duty scaffolds. For information not given in the appendix, the OSHA regulations mentioned should be consulted; I have not attempted to summarize them all.

While improper framing of a scaffold may cause it to collapse, the most common cause of injury to workers is inadequate planking and railings. Loose, defective, and too few planks are so commonly noted that I hardly need mention them, but anyone interested in construction safety should give the matter some serious attention. Regulations require the use of scaffold grade or equivalent lumber, as recognized by approved grading rules for the species of wood used. I did not recall much about grading rules, but on the assumption that people at the local lumber yards would have the information at their finger tips, I decided to pick their brains instead of doing a lot of research.

Stamford, Connecticut is a progressive city of more than 100,000 people in which there has recently been an unusual amount of construction as a result of urban renewal and influx of new corporate headquarters. I anticipated no problem locating scaffold grade planks, and all five lumber yards that I called carried 2 × 10 rough sawn spruce planks, full thickness, but none would sell them as scaffold grade material. One salesman mentioned that the firm had been warned by its insurance company against accepting orders for scaffold grade planks because they might be sued if an overloaded plank broke. All the yards were selling the planks; contractors had no convenient source of graded material.

I would like to make some recommendations for testing planks but, in view of the present tendency of any injured worker to sue anyone and everyone in sight, I would be very foolish to propose any testing procedure. In actual practice, I think the greatest danger is that planks that have been damaged will continue to be used, and only the workers themselves can effectively control this hazard. If a plank does not seem sound, I suggest it be immediately cut up so it can no longer be used. Scaffolds should be inspected every day, and should never be modified without the approval of a responsible supervisor.

Anyone interested in complying with OSHA regulations should read the sections dealing with railings on scaffolds very carefully; knowing the height and width, he will then be able to determine whether a standard guard rail is required. If it is not, he still has the problem of deciding whether it would be a good idea to install one. I sincerely hope that this section of the regulations will be revised to make it more acceptable as a means of avoiding falls on construction jobs. At the present time, even a casual observer can see that the regulations are being disregarded with impunity. Workers can be seen on scaffolds that have only a few scattered planks, some of which overlap their supports by several feet. Guard rails are very often omitted, and many that are erected lack the center rail or toe board. At the end of a shift, workers can be seen climbing down the poles of tubular metal scaffolds, using only the diagonal braces as footholds. Safety belts, lanyards, and lifelines are very often missing when workers are on suspended scaffolds where their use is very appropriately required by OSHA. Compliance officers can hardly avoid being aware of these conditions, but there seems to be a general reluctance to change those rules that are not actually wrong but that are just so conservative that relatively few contractors comply with them. In manufacturing plants, they are more often observed, but that does not prove that they should be part of the rules applicable to construction. My objection to them is that they encourage general violation of regulations, many of which should be strictly enforced. When reasonable rules are lacking and workers do as they please, they often take chances that result in falls. The statistics already cited prove that many lives and injuries could be saved if proper attention was given to the prevention of falls.

In England, I recently observed that scaffolds are usually erected by subcontractors who specialize in that work. They prefer tube and coupler scaffolding and the speed with which they work is obviously increased by their familiarity with this specialty. They bring plenty of planks, clamps, ladders, and pieces of tubing of many lengths to the job, so cutting is rarely needed. At most jobs that I saw, there was full planking at all working levels, and one plank set on edge and held tightly against the outboard plank served as a toe board. Tubing was used for rails. Ladders were usually wood, of very substantial construction, and were clamped in place. At the end of the day, contractors were apt to discourage children from climbing onto the scaffold by lashing a plank to the exposed face of every ladder, covering the rungs to a height of about 10 feet from the ground. Even at small jobs that would be done in America by men working from ladders, English contractors usually had scaffolding erected. At the end of the job, the scaffolder removed his equipment for reuse elsewhere. Some general contractors may do their own scaffold work, but the prevalence of signs indicating the names of different scaffolding contractors led me to believe most of the scaffolds were erected by specialists.

6.3.3 Ladder Safety

Mechanics fall from ladders so often that such accidents are widely accepted as inevitable. Still, it would be hard to find a construction worker who would admit that he is poorly informed about ladder safety. Installers of mechanical systems, accustomed to using relatively short ladders, are particularly apt to be overconfident. Of course, many who realize they are taking chances believe that if they don't get the work done that day they may be out of a job. Therefore they use whatever ladders are available, even when it is risky. For example, most mechanics know they should not stand on the top of a folding ladder, but if they have only an 8 foot ladder and they have to reach to a height of 15 feet, they take a chance.

Foresight is an essential part of good management; in anticipation of starting the job, competent supervisors consider not only the materials and manpower they will need, but the tools and ladders as well. Whenever I have represented the owner and been responsible for checking work, I have found that contractors who provided their workers with inadequate ladders also seemed willing to give us an inferior job if we allowed them to get away with it.

When the cost of a ladder is compared to the hourly rate for a construction worker, it is surprising that simple arithmetic would not be sufficient to induce contractors to buy the right ladders for their jobs, simply to save time. In addition, knowing full well that if a worker falls, the cost of the accident may be thousands of dollars, why do they send out rickety ladders? Ladder safety starts with straight thinking by responsible people.

Until the boss provides some good ladders, lecturing to workers about ladder safety is a waste of time. No amount of talk can mask insincerity; willingness to spend the money for good ladders is recognized as proof of good intentions. But sometimes the right policy is poorly executed; the wrong ladders are bought. Someone just calls up the local hardware store and orders whatever happens to be in stock. It may not be sturdy enough, the wrong type, or the wrong length. Too long is as bad as too short. Workers note that the boss seems sincere, but is a little stupid.

To create more positive attitudes, I suggest involving some experienced workers in the selection of new ladders. Get catalogs showing various types, some with antiskid and adjustable feet, and with different step, platform, and hinge arrangements. Include the models that have casters but that lock in place when anyone puts his weight on a step. For some jobs, look for accessories that will prevent falls if the worker loses his balance, and for brackets that will span window openings. When leading workers know more about ladders, they can influence others. An expert from a ladder manufacturer's staff may be willing to talk at a safety meeting. If possible, have him walk around the job before the meeting so he can comment on actual problems and be prepared to suggest ways to improve productivity as well as safety. By helping cut labor cost, he increases his employer's chances to make sales, and without some immediate financial incentive many contractors keep their safety engineers on a very tight budget.

OSHA rules and other follow-up topics can be handled at tool box meetings. See 29CFR 1926.450, covering both manufactured and job-made ladders. Ladder slope is a good starting topic. Every worker instinctively recognizes that there is a "best" angle at which to set up a straight ladder. OSHA rules say portable ladders *shall* be used at a pitch that corresponds to a horizontal distance from the top support to the foot of the ladder of *about* one quarter of the length of the ladder between the foot and the top support. How important is it to adhere to the prescribed slope? No author is likely to risk liability suits by saying what variation is permissible, and OSHA has not, as far as I know, defined the limits between which the slope may vary and still be within the "about" 4:1 ratio. With a short ladder and a couple of strong assistants to keep him from falling, a safety engineer can demonstrate that it is wise to keep close to the prescribed slope. If he sets up the ladder much steeper, it is necessary to "belly up" to it to keep it from falling, and any quick motion is likely to cause a fall.

OSHA also requires that "portable ladders in use shall be tied, blocked, or otherwise secured to prevent their being displaced." Note the use of the word "shall" again. It means "must." How many workers comply with this mandatory ruling? It is certainly violated more often than not, but Figure 6.16, a picture taken at the Limerick Generating Station being built for Philadelphia Electric by the Bechtel Power Corporation shows that at least one leading contractor abides by the rule. See also Appendix I for a close-up of a properly anchored job-built ladder.

In England, much more attention is given to ladder safety than in the United States. Scaffolders customarily lash or clamp ladders in place. Even when ladders are used instead of scaffolding, it is not unusual to observe them tied to pegs driven into the ground, using rope guys attached near the top of the ladder and run off at an angle sufficient to assure lateral stability. Where the ladder is apt to slip where it rests on the ground, it is sometimes tied to a stake. I am confident that if American workers want to anchor ladders securely in place, the task is not beyond their capability, and that they can devise ways to accomplish the necessary stability without spending much time doing it. On the other hand, if the craftsman does not want to tie off his ladder, no amount of instruction on how to do it will be effective; he will always find a way to prove that it is impractical, that it takes too much time, or that the ropes are in his way.

If the slope of a ladder is too low, the worker's weight puts a severe strain on the side rails. They may break, or the flexing of the rails may cause the worker to fall to one side, carrying the ladder over with him. It is usually well within the capability of workers or foremen to solve access problems of this kind, but failure to take the trouble to do so causes unnecessary accidents.

Tool box meetings are also a good time to remind workers of causes of accidents that fall into the general category of failure to use common sense. For example, if a worker sets up his ladder near a door and someone opens the door, one or both may be hurt. On uneven ground, if a worker uses a brick under one leg of the ladder and it slips out while he is on the ladder, the ladder will topple over. When two or more workers try to use the same ladder at the same time, someone's hand is likely to be stepped on, the ladder may break under the excessive load, or the top man may drop something on the one below. The normal procedure of facing the

Figure 6.16 A portable ladder may be tied off in several ways. This one has been lashed to the hand rail at the top and a rope has been run from one of the lower rungs to the concrete pedestal on top of which the men are working. (Photo courtesy of Philadelphia Electric Co.)

ladder when climbing should seem so natural that it is hard to understand why anyone would come down with the seat of his pants against the ladder. But they do it, and if a heel catches, a fall is unavoidable. Ladders may be seen resting on a plank that spans the distance between two steel drums, certainly an invitation for an accident. While adjusting extension ladders, some workers put their hands on the rungs, inviting painful injury if the rope breaks or the catches fail to operate properly. Sliding down a ladder is fun, provided one doesn't get splinters or fall, but having fun is not part of the job. Stunting and horseplay often backfire.

Accidents often occur when it is necessary to rest a straight ladder against a round column or a vertical pipe. If the ladder is not too long and flexible, it may be safe for one worker to climb up and tie the top rung securely to the pipe or column, while a helper steadies the bottom, but there is always the chance the ladder may slip or twist while this is being done. Clamping or bolting a crosspiece to the top of the ladder with a curved or vee notch to prevent side-to-side slippage helps, but there is still the danger that the ladder may twist around. Workers should be urged to avoid this hazard whenever possible by planning other means of reaching the work to be done. It may be possible to use a folding ladder, or to set up a few pieces of scaffolding. Each foreman and safety supervisor should give serious consideration to this hazard, and try to minimize it by whatever means are available to them.

When work must be done on high ladders and, particularly when the weather is windy and the job requires that both hands be used, workers should be provided with safety belts, fall arresting devices, or both. When proprietary fall arresting equipment is used, adjust it so the maximum free fall is as short as possible. When a fall is stopped suddenly, injury can result. Manufacturers of such equipment offer shock-reducing gear.

Metal ladders are unsuitable for use when contact with live electrical circuits is possible. Electrocution is a hazard to which mechanics in many trades may be exposed; it is by no means limited to electricians. Use wooden or fiberglass ladders whenever there is electrical equipment within reach.

If a ladder has been damaged, tag it "Danger-do not use" or some equivalent warning, and get it to a safe place where workers will not be tempted to use it. If it is not worth repairing, cut it up so someone will not salvage it and get hurt, or attempt a makeshift repair.

Good ladders are not cheap; insist that workers take care to avoid snapping the side rails when lowering them to the ground. If the ladder is too long for one worker to keep under full control, he should always have a helper to avoid injury to himself, damage to the ladder, or both. Ladders are not suitable substitutes for bridge planks or scaffolds. Don't permit them to be used in a horizontal position.

To keep disputes regarding ownership of ladders from consuming valuable time, I recommend that every ladder be stenciled several times with the owner's name. In addition, some spots of bright color at the ends may help its owner find a ladder that has been "borrowed" without permission. Wooden ladders will last longer if given a coat of wood preservative and spar varnish, but paint should not be used; it can conceal cracks.

Give some thought to how ladders should be left at the end of the day. If the site is not enclosed, unwanted visitors may be encouraged to climb them and possibly get hurt. It could be claimed that leaving the ladder constituted an attractive nuisance and made its owner responsible for the intruder's injuries. In some neighborhoods, it is advisable to lock ladders to keep them from being stolen. When several subcontractors are on the job, it is not unknown for good ladders to disappear when one of them leaves the site. I can thus understand the attitude of contractors who buy only the cheapest ladders, reasoning that anything worth stealing will be taken. From a safety viewpoint, this policy is to be discouraged and it may require the attention of the general contractor to keep the subs in line.

6.4 Fire Damage Prevention and Control

There are three principal ways to minimize fire damage:

1. Make the structure as nearly fire-safe as possible (see Section 6.4.1).
2. Avoid or control activities that can start fires (see Section 6.4.2).
3. Be prepared to extinguish small fires before they get out of control (see Section 6.4.3).

The foregoing apply both during construction and after completion, but my primary topic is construction safety and, therefore, I have mentioned hazards in occupied buildings only when they exist prior to completion and are not caused by the tenants or other occupants of the building after it has been accepted by the owner.

6.4.1 Fire Resistance During Construction

Fire codes require many features that do not protect construction workers. Sprinkler systems, smoke alarms, fire doors, and pressurized stairwells are not operable until most buildings are nearly complete. Consequently, the construction safety engineer needs to consider fire safety from a different viewpoint than that of the design professional. He must try to keep the structure as fire-safe as possible at every stage of construction. In so doing, he will also assure the owner that the completed structure is as safe as intended.

The MGM Grand Hotel fire in Las Vegas in November 1980 caused 85 deaths. Sixty one were killed on the sixteenth and higher floors by smoke that spread rapidly from the fire on the lower few floors. The smoke could have spread with equal or greater rapidity if the fire had occurred while the building was still under construction. Fire broke through the bottom of a stairwell, filling it with smoke. Flames also got through doors into elevator shafts, and went up unsealed pipe chases. Airconditioning ducts had smoke dampers that were inoperable. Parts of the airconditioning, heating, and ventilating system did not shut down during the fire. Heat melted plastic pipe that passed through fire walls and allowed flames to follow.

As work proceeds on any job, care should be taken to block all openings through which smoke or heat could pass from floor to floor or through fire walls. If design deficiencies become apparent during construction, neither the owner nor the architect may appreciate having them brought up for discussion and of course there will be the inevitable debate about who should pay to correct them. The easiest course of action is to say nothing but if there is a fire and investigators observe that the contractor knew or should have known about hazards and failed to correct them or to notify the owner, he may have a difficult negligence suit on his hands.

Material selection is a controversial subject. Tests do not always give consistent results, and this makes it difficult to predict how a material will behave in a fire. In fact, the whole science of predicting how a fire will develop in any given space is in its infancy. Some authorities believe that there is a great advantage to the use of natural fibers instead of synthetic materials; others disagree and think the differences are of minor importance in comparison to other factors that influence the extent of fire and smoke damage. Some manufacturers add fire retardants to their products and, when this adds to the selling price, they must assume the burden of convincing designers to insist that all approved alternative products have equivalent fire-retardant ingredients. Here we have the "or equal" problem as a deterrent to the use of safer materials.

Materials should only be installed in ways approved by their manufacturers. Combustible insulating materials that are intended for use in walls between fire-resistant barriers should never be used when they are exposed or protected only by panels that are themselves readily damaged by fire. Many of the synthetic foamed plastic insulation boards will decompose readily and release toxic gases. Keep such materials out of the building until workers are ready to use them and to apply the appropriate protective surfacing. Do not assume that materials of the same general nature are necessarily the same. Check test reports. For example, polystyrene insulating materials may be foamed by agents that include either methylene chloride or hexane. The behavior of the product can be different, depending upon which foaming agent is used.

Progressive architects and owners have begun to recognize that sprinklers may be needed above as well as below suspended ceilings. There is renewed interest in plaster as a decorative material. When costs are comparable, preference should be given to the more fire-resistant material. Some progress is being made, but it seems to be less than adequate.

Contractors should be wary of installing any material in an unusual way. For example, oriental rugs are sometimes displayed as decorative wall hangings, and in this position they are much more flammable than when lying flat on a floor. Similarly, a photomural will not burn nearly as fast when pasted on a plaster wall as when mounted on fiberboard and framed.

Whenever possible, give priority to the installation of sprinkler systems, smoke alarms, automatic dampers, and any similar items that reduce the possibility of smoke or fire damage. As a building nears completion, the opportunities for a fire to get started can increase, particularly when the final touches are being added and furnishings are being delivered.

Cartons, wrappings, and crates add to the hazard of fire. Before this type of activity begins, the fire protection systems should be fully operable.

6.4.2 Fire Prevention

Regardless of how much care is given to the design and construction of a building in order to make it fireproof, there will always be some flammable materials in it. Even a prison can have a disaster as evidenced by the fire in Biloxi, Mississippi, in November 1982 where inmates were asphyxiated by poisonous gases from a burning foamed plastic mattress.

Many workers cannot get through the day without smoking; some will do it in spite of all rules to the contrary. It is safer to designate areas where smoking is permitted than to attempt to enforce no smoking rules. Equip those areas with sand buckets for butts and fire extinguishers. Make it easy for people to cooperate.

Constant vigilance is needed to keep fires from being started, not only as a result of smoking, but also as a result of defective electrical cords or tools, welding, salamanders, and even lightning or static electricity. During construction, products may be used that are more easily ignited than any commonly present in a completed building.

People normally recognize materials that will burn, but they seldom differentiate clearly between the degrees of flammability that they may encounter. For example, there are important technical differences between flammable and combustible liquids, and yet in common usage the terms are often interchanged.

Every liquid that will burn has a flash point; if it is heated to that temperature, the vapor just above the liquid can be ignited. Several testing methods have been devised, and regulations regarding the handling and storage of flammable or combustible liquids normally specify the test method that shall be used to determine the flash point. In view of the differences between flash points of the same liquid as determined by different test methods, it is important to record the test method used.

The U.S. Department of Transportation (DOT) classifies a liquid as a "flammable liquid" if its flash point is under 100°F as determined by one of the test methods approved in 49 CFR 173.115. If the flash point is 100°F or over, but under 200°, the DOT classification is "combustible liquid". If the flash point is 200° or above, the DOT does not require it to be classified as a combustible liquid for purposes of transportation. On the other hand, the National Fire Protection Association considers all liquids that have flash points of 100° or more to be combustible liquids. They recognize three subclasses of flammable liquids and three subclasses of combustible liquids.

While these definitions of flammable and combustible liquids differ in certain respects, they all classify liquids having flash points below 100°F as flammable and those with higher flash points as combustible. The distinctions are important because they affect the storage and shipping methods that may be permitted. I was therefore surprised to find the OSHA fire prevention regulations use 140°F (60°C) as the flash point below which a liquid is considered a "flammable liquid". It is classified as a "combustible liquid" if the flash point is 140°F or above and less than 200°F. See 29CFR 1926.155(c) and (h). Safety regulations are apt to be difficult enough to follow without adding to the problem by changing the definitions of terms. I hope that the various agencies interested in fire prevention will get together and agree on basics; it would make it easier for all those who are attempting to comply with their regulations.

Local fire departments may have regulations that differ from all the foregoing and, in particular, they may require use of a test method that is not listed by DOT or NFPA. Therefore, if any flammable or combustible liquids must be stored at a job, it is important to discuss safety measures with the local fire department and abide by their rules.

Always handle gasoline with special care; have no more of it around than is absolutely necessary, keep it in safety cans made specifically for the purpose, and store the cans in a manner approved by the fire department. Naptha and other highly flammable liquids merit similar handling.

When filling gasoline tanks of equipment from portable cans, be sure the spout of the can is bonded to the machine because the flowing liquid can create static electricity that can produce a spark and ignite the vapors if the can and the tank are isolated while the liquid is flowing. This precaution is automatically assured when a car fills up at a gas station; the hose always has a metal spout and the hose itself has a conductive braid that is grounded.

Use high flash point or nonflammable solvents for cleanup. Never use gasoline for this purpose. It is too easy to ignite, and it often contains toxic additives. Use gasoline as a fuel for engines and for *no* other purpose. *None.*

Of the nonflammable solvents, strictly avoid the use of carbon tetrachloride. For generations it was a common household spot remover and was even used in fire extinguishers. It was very popular with electricians who had occasion to clean electrical equipment. Before it was widely publicized that carbon tetrachloride vapors were highly toxic, one of the men working under my general supervision was found dead in a confined space where he had been cleaning a connection box. I also had a close call when I tried to put out a small fire with a carbon tetrachloride fire extinguisher. The chemical decomposes under high temperature and produces extremely irritating and toxic gases.

Other flammable materials on typical jobs include salt hay, wood, many plastics, cloth, oils, and thinners. Be particularly cautious about the use of treated materials; some saturants or coatings are very easily ignited, whereas others are fire retardant. Tarps used near salamanders should always be of a fire-resistant material.

Synthetic materials that do not necessarily burn too readily can decompose chemically when exposed to flame. Polyvinyl chloride (PVC) has this characteristic, and the fumes are dangerous; they include carbon monoxide and hydrogen chloride. PVC is a very useful and versatile product, often sold in tubular and sheet form. It is also used to manufacture molded products such as pipe fittings and insulation. In 1975 a fire at the New York Telephone Company's main switching center destroyed wires that were insulated with PVC and 239 firefighters were overcome by the toxic fumes. Recently, New York City has discontinued use of PVC conduit for wiring in subway stations.

Ignition of a flammable material may occur spontaneously without direct contact with a spark or flame. This characteristic of oily rags is well known, but fires are still started by piles of painter's drop cloths. Oxidation of the oils takes place slowly, but heat is released notwithstanding and when the temperature gets high enough ignition will occur. When welder's oxygen comes in contact with an oily material, the rate of oxidation can be explosively rapid.

Under some conditions a fire may smoulder for hours before bursting into flames. Extensive tests have shown that a cigarette butt can ignite fibrous material hours after it would normally have burned itself out in an ash tray. These delayed ignition fires are particularly dangerous because all the workers may have left the site, leaving nobody there to fight the fire before it gets out of control. Even if a watchman happens to be on hand, it may be impossible for him to put it out before being overcome.

In August, 1981 fire broke out about 15 minutes after construction workers left for the day and flames shot up the face of the partially completed building. They set fire to scaffolding planks and forms. Then the heat melted aluminum parts of the scaffolding and damaged a crane so severely that a 100 foot jib and parts of a 350 foot boom had to be replaced. There was also damage to the cables of the crane. The probable cause of the fire was a discarded cigarette. Fortunately there were no injuries.

Once a fire starts burning it can spread with surprising speed. This makes it imperative that someone be watching whenever there is a possibility that a fire may be started by soldering, welding, or torch burning. The intense light of the torch or arc plus the worker's dark goggles make it impossible for him to see a fire until it has gotten well started. A watcher with a fire extinguisher should therefore assist the welder whenever he is near flammable materials.

Welder's goggles have been mentioned in Sections 3.5.2 and Appendix D. Inexperienced workers should be warned never to watch a welder without wearing protective goggles. Not only is eye damage possible, but a person who is momentarily blinded can trip over an obstruction or stumble into a danger that would otherwise be easily avoided. Slag from metal cutting operations can also cause burns and falls. The first time a hot chip fell into my open shirt collar I learned to keep it buttoned while near any job that produced a shower of hot material. Who can say what will follow if a hot piece of slag gets into someone's shoe, glove, or shirt.

When a burning torch is being used to cut steel, the metal should never rest on a rock or concrete surface in a position that will let the flame impinge directly on the masonry. Intense heat can cause some rocks and aggregates to pop or spall off with considerable force. I narrowly escaped being hit by a sharp-edged piece of rock the size of a dinner plate and about an inch thick that flew from a boulder I was splitting by heating it with a kerosene flame thrower. The process of alternating heat and chilling with cold water was effective on most of the boulders, but one was different; nothing happened until there was a loud explosive noise and the slab flew about 30 feet through the air.

Oxygen and acetylene cylinders must never be used as rollers. Close the valves and put on the valve protection caps when moving oxygen tanks; remove the keys when handling acetylene cylinders. Fasten tanks securely while they are in use so that if anyone trips over a hose it will not tip them over and possibly snap off a connection. There are a number of specific rules governing welding operations, and every one of them is important. I therefore recommend that 29CFR 1926.350 through 1926.354 be examined in detail before welders and their equipment arrive at the job. At that time, try to anticipate conditions that may exist at various stages of the job, and make notes to follow up on items that will require attention later. For example, the first welding work may be done in the open air where it is obvious that no ventilation problem exists, but later on it may be necessary to weld piping in machine rooms and, if the operation will create fumes containing lead, zinc, cadmium, or chromium, additional general or local exhaust equipment may be needed.

Accidents have been caused by unauthorized use of a welder's equipment. Oxygen and compressed air are inherently different materials. If oxygen is used to pressure test a pipe or vessel that contains oily residue, it can cause an explosion. The gas in a cylinder may be under extremely high pressure, 2000 psi or more. Using this pressure to blow dirt off anything may cause small parts to fly through the air like bullets.

Acetylene forms an explosive mixture when present in the air in concentrations of between 2.5 and 81% by volume. Note the very wide range between the lower and upper explosive limits. Gasoline vapor is explosive only when present in concentrations of 1.4 to 7.6% and while I do not want to minimize in any way the dangers associated with spillage of gasoline, I do want to emphasize the characteristics of acetylene that make it particularly hazardous. It is not generally known that acetylene will react with copper to form an explosive compound in the presence of moisture. Copper and copper alloy parts of acetylene handling equipment are made of alloys that have been formulated to avoid this hazard, but ordinary copper tubing has not. Therefore, it is dangerous to use copper tubing for connections to acetylene tanks or for manifolds.

Acetylene cylinders contain a porous material saturated with acetone, which is a flammable liquid. The acetylene gas dissolves readily in the acetone, and it is this property of the gas that makes it possible to transport it under pressure in cylinders. Never attempt to compress acetylene or to refill cylinders; that may be done safely only where specially trained and equipped workers are available. Manufacturers recommend that the acetylene be drawn from a cylinder at a rate no faster than that which would exhaust the cylinder in about seven hours. For heavy burning, it is therefore necessary to manifold cylinders. Follow the manufacturer's recommendations regarding the connections to the manifold and the use of regulators. Keep reserve stocks of oxygen and acetylene cylinders separate and, if any leaks are found when testing connections, correct them promptly. If a leak cannot be eliminated by simply tightening connections, move the cylinder to a safe spot well protected from any source of ignition, crack the valve slightly, and allow the gas to escape. Be sure there is free movement of air around it, so that no gas pockets can form. Tag the cylinder to indicate where the leak is to be found.

Arc welding cables get hard use; their insulation should be checked frequently to be sure a fire cannot be started by an accidental short between the cable and some grounded part of the structure.

6.4.3 Fire Control

If a fire starts, sound an alarm to the fire department even though it may seem that there will be no difficulty putting it out with extinguishers or hoses already at hand. Anybody can underestimate the difficulty of controlling a blaze and, if it proves to be a case of arson, several more fires may be started while the first one is claiming everybody's attention.

If the fire is out by the time the fire trucks arrive, it is still important to have professional firefighters check to be sure no smouldering embers remain. They may also want to investigate the cause of the fire.

While firemen are on the way to the job any firefighting that is attempted should be concentrated where it will reduce the risk of injury to workers and then to the point where the greatest damage to the structure may occur. This latter location may not be immediately apparent.

In 1981 an arsonist set fire to a wooden control house at one end of the 220 foot long lift span of a bridge across the Mississippi River. The heat caused cables supporting the counterweights to snap, allowing the span to fall into the river. It had to be demolished. I cannot say whether concentrating fire-fighting streams on the structural steel and the cables could have prevented the collapse, but in retrospect it appears that the wooden control shed would have been a loss in any event and that it might well have been better to try to keep the steel parts as cool as possible.

In another arson incident during the summer of 1981, fire in a warehouse under the 62nd Street Bridge over the Allegheny River in Pittsburgh overheated the exposed steel bridge girders. Their weight and that of the concrete deck caused the span to sag about six feet. Damage to the warehouse was about $250,000; the cost of bridge repairs was three times that amount.

These incidents illustrate the vulnerability of structural steel to fire and, as anyone who has done work at a blacksmith's forge will recall, steel that was heated red hot initially still retains enough heat to be readily bent even after the color has disappeared. Structural members can lose over half their strength before they show any trace of color, and by the time they are at a temperature corresponding to a dull red, their strength is probably well below 10% of normal.

The behavior of aluminum exposed to heat is even more spectacular. If it is raised to a temperature at which steel would just begin to show a red glow, aluminum will suddenly melt. It has practically no plastic range; it is apparently intact one moment and molten the next.

Heavily built wooden structures often resist fire better than steel ones. It takes much longer for a 12 × 12 wooden beam to fail when exposed to fire than a steel member of equivalent strength. To prevent structural failure, it is therefore important to focus attention on those members that will fail first.

If the structure has been enclosed, as the temperature in the space rises, everything that will burn suddenly ignites, often with explosive intensity. This dangerous phenomenon is called "flashover," and workers who might be tempted to try to control a small fire should be warned of this danger. A fire can get out of control so suddenly that failure to recognize the danger of flashover can be fatal.

The ability of fires to spread rapidly after having smouldered for hours emphasizes the importance of good housekeeping to eliminate piles of rubbish in which a fire may be undetected before it flares up. When it does become visible, fast action will be needed because the mass of hot material under the flame may be ready to support a large fire, out of all proportion to what might be expected.

In order to be of any value, portable fire extinguishers must be of adequate size and suitable type. Three classes of fires should be anticipated at typical jobs:

> *Class A* fires are the most common and are fueled by paper, salt hay, wood, and other common combustible waste material. Water is generally preferred to put out such fires; it has excellent ability to cool flammable rubbish and, while other extinguishing agents will do the job, they are usually less effective, pound for pound.

> *Class B* fires are fueled by oil, gasoline, grease, paint, and solvents. A solid stream of water may spread a fire of this type, and dry chemical, carbon dioxide, or multipurpose extinguishers must be used.

Class C fires occur in or around electrical equipment; therefore the extinguishing agent should not conduct electricity. Dry chemical and carbon dioxide extinguishers are often used; they are safe and do not usually damage electrical equipment. Water should not be used unless and until it has been positively determined that the power is off, and the electrical gear is already ruined.

Extinguishers suitable for use on different classes of fires are marked by one or more standard symbols:

Class A extinguishers are marked by the letter A in a green triangle. They are marked "for ordinary combustibles."

Class B extinguishers are marked with the letter B in a red square with the words "flammable liquids" near the symbol.

Class C extinguishers have the letter C in a blue circle with the words "electrical equipment" adjacent to the symbol.

Occasionally the letter D on a yellow star may be seen on an extinguisher; it is intended for use where combustible metals are used, and is unlikely to be of interest to contractors.

Many extinguishers are marked with two symbols, but that does not necessarily mean that they are the best selection; versatility may have been obtained at the expense of capacity. I suggest the manufacturer's literature be carefully studied before a selection is made.

Every worker should be shown how to use the extinguishers available to him and should know where to find them in an emergency. Until work has progressed to a point where they may be hung at marked locations, certain people may have to be assigned responsibility for fire safety, and this job would naturally continue whenever welding or soldering was being done.

6.5 Roofing

Employer's Insurance of Wausau studied compensation claims totaling over $5 million paid by them to 1668 roofers, an average of over $3000 per accident. Worker's compensation classification 5551, roofers, is one of the three or four highest premium classifications in the construction field. Typical rates add 20 to 30% to the labor cost.

Wausau found that claims resulting from falls cost about 33% of the total, and materials-handling accidents accounted for about 20%. About 15% of the payments were to burn victims. Materials-handling injuries involved the worker's backs in 80% of those cases.

The National Institution for Occupational Safety and Health (NIOSH) publication no. 78-194, *Health and Safety Guide for the Commercial Roofing Industry* is available from the Superintendent of Documents, Government Printing Office, Washington, D.C. 20402. It deals principally with safety hazards to which workers on built-up roofing jobs are exposed, and was prepared in cooperation with contractors, OSHA, unions, industrial hygienists, and a number of other groups. It includes many on-site pictures and is well worth its nominal cost.

NIOSH and insurance companies have safety specialists, some of whom are available to contractors who need help in identifying on-site hazards and possible practical solutions.

Roofing is a labor-intensive craft, and yet less than 15% of the contractors have more than 20 employees. The small size of the typical firm and the vagaries of the weather make it impossible to avoid doing jobs when they must be rushed, and this is a major underlying cause of accidents. I suspect that if contractors gave appropriate weight to the potential costs of accidents, they would find it easier to justify the purchase of more materials-handling equipment and better maintenance of kettles, cutters, and other tools. Anything that will make it possible to speed up the work while the weather is fair will take some of the pressure off the foremen and workers and should help reduce accidents.

In typical small organizations, the supervisor is actively involved in the actual work, or is moving about from site to site. In either case, he must often rely on workers to avoid dangers without being watched or warned. This makes training important, and the instruction must include enough "selling" to make the men want to follow the best practices without anyone standing over them. This training can be given to permanent employees, but it is very difficult to indoctrinate the temporary help hired for one or two big jobs. The regular workers must be relied on to keep the others out of trouble.

The foregoing considerations should make it clear that jobs need to be planned in anticipation of possible dangers. For example, if a cutter has to be raised to a roof, the men who will have to do it should know in advance how to go about it, and should have all the necessary equipment available. When everybody is ready to get the job started, it is too late to figure out how to proceed, and under such conditions men make do with whatever is handy. They also tend to get the job started without figuring out how to take the next step; they may get the tool up to roof level and be unable to get it off the skids and onto the roof without taking chances. Someone may get badly hurt in the process. Talking about the next day's work pays off; I have often thought I had a job well planned but before morning I became aware of a possible snag that had to be eliminated.

I realize that these generalizations may seem too vague to be of much practical application, but I have found that most workmen have more common sense and more good ideas than some foremen expect of them. In a small organization, it is important to involve all the people and make the best possible use of their talents. I might list many specific precautions, such as avoiding burns by proper charging of kettles and by keeping a close watch on the temperature of the bitumen, but in the long run, the specifics are already known to the men whereas the accidents are caused by failure to anticipate the unexpected.

But chemical hazards need to be determined by supervisors; workers cannot be expected to read and heed warning labels on containers of solvents or cements. They also have to be supplied with safe materials for cleanup or they will use whatever is handy, including highly hazardous gasoline. Workers also tend to underestimate the danger from vapors and fumes; they recognize the immediate discomforts but ignore the potential dangers inherent in the use of products known to cause cancers or organic injury. Although they can be encouraged to let the wind work with them, staying upwind of dusty or smelly activity, this is sometimes impossible and respirators are needed. Management is responsible for providing suitable safety gear and for insisting that it be used when needed.

On every job, someone must have the authority to stop work when in his judgment it is unsafe to proceed. With loose supervision, nobody is apt to risk criticism by stopping before the job is completed, particularly if there is a chance that a threatening storm may pass over. Consequently, men keep working when they should be preparing for wet weather, icing conditions, or high winds. If a foreman makes a mistake and secures the job in anticipation of dangerous weather conditions that fail to materialize, he deserves praise rather than censure. Any fool can criticize him for making a mistake when only hindsight is needed, but it takes a good man to take action when he knows he may be in trouble with the boss for doing what he thinks is necessary for the safety of the men.

Supervisors should keep in mind their responsibility for the safety of their men, and this includes watching for any indication that someone may be reaching the limit of his endurance on a day when the work is hard, the weather is difficult, and there have been few chances for coffee breaks.

In spite of the poor accident record of the roofing industry, OSHA has failed to make it easy to find the applicable regulations. In 29CFR 1926, rules applicable to some trades are easy to find: subpart J deals with welding, subpart N is devoted to cranes and similar equipment, steel erection is the topic of subpart R, and there are several other similarly well-organized subparts. But there is no comparable subpart about roofing. Anyone interested in the applicable regulations must recognize the hazards and look them up wherever they may be found. Consider gasoline, for example. There is no entry in the OSHA manual index under "gasoline," none under "solvents," and nothing about "cleaning solutions." If one decided to look in the index under "fuels," he would find references to subpart F on fire protection and prevention, but I have failed to find anything in that subpart about the danger of using gasoline as a cleanup solvent.

Regulations applicable to roofing brackets are in 29CFR 1926.451(u) which is part of subpart L on ladders and scaffolding. Tucked away under the heading of roofing brackets is the requirement that a catch platform shall be installed below the working area of roofs more than 16 feet from the ground to eaves, if the slope is greater than 4 inches in 12 inches and there is no parapet. This platform is required to have a standard guardrail, midrail, and toe board. Finally, the paragraph permits the platform to be omitted if workers have safety belts attached to a lifeline. Anyone interested in chicken ladders will find them indexed as crawling boards, paragraph 1926.451(v). That section tells how they shall be made and requires that a lifeline be strung next to each crawling board to serve as a handhold, but it does not say that a safety belt must be worn and attached to the lifeline.

Those who have subscribed to volume III of the OSHA subscription service, which updates regulations on construction standards and interpretations, will have received new rules for guarding low-pitched roof perimeters during the performance of built-up roofing work, 29CFR 1926.500(g) to 1926.502. Three options are available:

1. A motion stopping safety system (MSS system) may be used, consisting, by definition, of standard railings, scaffolds or platforms with guardrails, safety nets, or safety belt systems.
2. A warning line system to rope off roof areas closer than specified distances from the roof edge, plus a MSS system for workers using mechanical equipment between the warning lines and the edge of the roof. This MSS system requirement is waived if a safety monitoring system is used, and no mechancial equipment is used between the warning lines and the edge of the roof. A safety monitoring system requires that a "competent person" must monitor the crew and warn workers who are unaware of a hazard or working in an unsafe manner. He must be on the same roof, and close enough to the workers to communicate with them verbally at all times.
3. On roofs 50 feet or less in width, a safety monitoring system may be used without the warning lines. I have just very briefly sketched these rules; they must be read to get all the details, which include five pages of examples to show how to tell whether the roof is 50 feet wide or not.

Perhaps existing regulations are excessive and possibly they should leave it up to contractors to recognize more hazards themselves. Certainly anyone who plans to bring heavy equipment onto a roof should first check the deck for weak spots; I don't consider it the function of a safety engineer to assume responsibilities of that kind, even though safety is involved. It is difficult to determine what is good construction practice and what more is necessary for worker safety; in the roofing business they are very closely related.

6.6 Avoiding Structural Collapse

Most safety effort is directed toward the prevention of injuries to individual workers, but if the whole structure collapses it is a safety hazard to everyone. In many instances, structural failure is caused by the actions of steelworkers, riggers, or others obviously involved with the strength and stability of the building, but in some instances material handling by subcontractors, cutting and patching, vagaries of the weather, or seismic action may be involved. To deal adequately with the topic is beyond the scope of this book, and readers will find Jacob Feld's *Construction Failure* of value; it is another volume in the Wiley Series of Practical Construction Guides. My comments from a safety viewpoint are concerned with the indirect causes of collapse rather than the technical details. I must necessarily mention some mechanical deficiencies, but merely as illustrations of how failure to coordinate the activities of people can lead to trouble.

At the October 1981 convention of the American Society of Civil Engineers, the theme of one session was Safety Considerations During Construction. W.F. Chen* and X.L. Liu† of Purdue University presented "A Study of Concrete Framed Structures During Construction" in which they pointed out the need for better ways to determine when it was safe to remove forms and shoring. At present, much reliance must be placed on the judgment of the contractor who has an urgent need to proceed as rapidly as possible and who has only general guidelines to supplement his experience. Structural analysis of the building while it consists of some freshly placed concrete and some partially cured material, all interrelated by shores and forms, would be an engineering nightmare, yet whenever forms must be stripped someone must make a decision that involves the safety of everyone on the job. Chen and Liu are working on the application of computer technology to this complex problem.

At the same convention, "A Reliability Analysis of the Tier Method of Construction of Steel Buildings" was submitted by S. Karshenas** and A.H.S. Ang.‡ They pointed out the critical importance of wind loading during erection of steel frame structures and calculated that the mathematical probability of failure at different stages of construction of a typical 10 story building, built in Chicago during the months of July and August, could range from less than 10^{-7} to as high as 10^{-2}. Shear wall frames were found to be particularly high risk structures between the time that the steel was erected and the construction of the shear wall.

These papers illustrated the importance of technical analysis of structural stability during construction and the need for someone to be on the

*Professor of Structural Engineering, Purdue University, Lafayette, Indiana.
†Graduate Student, Department of Civil Engineering, Purdue University, Lafayette, Indiana.
**S. Karshenas is Assistant Professor of Civil Engineering, Marquette University, Milwaukee, Wisconsin.
‡A.H.S. Ang is Professor of Civil Engineering, University of Illinois, Urbana, Illinois.

job who is qualified to make structural engineering decisions was emphasized by A.P. Arndt, Director of Engineering, American Bridge Division, United States Steel Corporation, Pittsburgh, Pennsylvania, in his paper "Structural Safety During Erection Operations—A Contractor's View." He commented on the American Institute of Steel Construction's Code of Standard Practice as it applies to erection work and observed that dangers could result from division of engineering responsibility. He pointed out that when contractors are involved in work where standard procedures are not well defined and understood by all concerned, daily planning meetings are advisable to assure that individuals directing the work understand the intent of the procedures and how their individual activities contribute to accomplishing the desired result. He warned about overextended use of temporary connections and overloading of the structure during materials-handling operations.

Owners may not be aware of the complexity of erection problems, and the professionals best qualified to supply the technical guidance to contractors may be hesitant to urge owners to include on-site work in their consulting agreements. They cannot voluntarily assume additional duties, yet they may consider it unprofessional to imply that contractors may not be fully qualified to make the critical decisions needed as work proceeds. I hope that more owners will recognize that it is in their best interests to be certain there is adequate technical talent on the job and that responsibilities are clearly defined.

Consider the case of the 11 workers who were killed and the 23 others who were injured when a five story flat slab concrete structure collapsed at Cocoa Beach, Florida in 1981. The roof slab was being poured when one column punched through the top floor slab and in a few moments only the stairwell of the 58 × 242 foot building remained erect. In the following eight months the State of Florida Department of Professional Regulation, the U.S. Bureau of Standards, and OSHA investigated and agreed that collapse was due primarily to serious design deficiencies that had not been detected during the customary plan review procedure. The structural design calculations had been made by two retired NASA engineers, one of whom had signed and sealed all the 79 pages even though he later admitted he had made only a cursory examination of all but the one page he prepared himself. The municipal engineer acknowledged that he had merely skimmed through the calculations before approving the plans. The two structural engineers made field inspections during construction and testified that prior to the collapse they had examined cracks around columns on the second and third floors but did not consider them significant.

The designers and the plan examiner were all clearly at fault, but investigators also criticized the contractor for poor placement of steel, slabs that were less than the specified thickness, columns that were smaller than called for in the design, premature stripping of flying forms, and inadequate shoring.

OSHA cited the general contractor for six serious violations, and several subcontractors also received serious citations. The engineer who made most of the design calculations surrendered his license but admitted no wrongdoing. The widow of one of the deceased workers entered suit against several parties other than his employer. I assume that after the collapse, the owner recognized that someone should have insisted on fully qualified technical supervision of the work, but I have not observed any industry-wide movement in that direction having been reported.

Checking design calculations is very difficult; a person who makes an error is likely to overlook it when checking his own work, and it is surprising that the currently routine examinations by plan examiners do not overlook more errors. It is unreasonable to expect them to check complex calculations made by computers without having technical skills and equipment equal to those of the leading professional designers. Yet the position of plan examiner in most communities is unlikely to carry an adequate salary to attract the most highly qualified engineers. I think this problem of adequate checking can only be resolved by the engineering profession; legislation and governmental regulation has gone just about as far as I think is appropriate, and it would be unfortunate if public reaction to structural failures were to hamstring the industry by ill-conceived legislation that increased costs without solving the problem.

Even after a project has been completed and accepted by the owner, those who designed and built it continue to be very deeply involved if a catastrophic failure occurs. Consequently, on-the-job safety problems during construction must not be permitted to overshadow the broader long-term view that public safety must always be remembered. Whatever risks construction workers take and their employers neglect to control involve their relationship to each other, but if the structure they build is unsafe, it affects everybody who may pass through its doors. People in responsible positions should occasionally consider the possibility of serious trouble, and reflect on the extent to which they might be personally considered negligent. I have already mentioned some limitations to the control individuals may have over errors, but I have not stressed some things that could be done to make our system work better.

Improved communication would help. After an accident, people who could have prevented it may testify that they were not aware of a change that should have been brought to their attention. I know this is not easy to do; important assignments seldom are. But it is worth the trouble to be sure people know what is going on if there is a possibility they might want to change the course of events.

There is always pressure on plan examiners and designers to expedite approvals. I have seldom heard of an equivalent amount of pressure being applied to assure that the reviews have been thoroughly made. There is never enough time to do important work with complete thoroughness while construction is being delayed, but there is always plenty of time to find the smallest oversights of the reviewer after a building collapses.

Minor changes that do not require submission to a plan reviewer or designer are difficult to separate from "important" changes. There is more pressure to omit submission than to pass the change along for review. Yet some seemingly minor items have caused major catastrophies. The decision to request approval or proceed without it should be made by people with mature judgment. Experience tends to make people more cautious, more aware of their own fallability, even aware of their own limitations in technical fields where others are experts. Priority attention should be given to the selection of people to whom authority to route problems will be delegated.

Owners should recognize the remote possibility that they are making a mistake by accepting the lowest bids. Many seem to have an unshakable degree of confidence in the so-called competitive bidding system and are not sophisticated enough buyers to know when they are being taken. They might do well to find out more about the outfits from which they accept bids. It can easily cost more to force a contractor to do a job properly than to pay the differential between the highest and lowest bids. How much would it cost to be absolutely certain that every precast beam and every subassembly was up to specifications? It is easy to get away with deception and costly to root it out.

Nineteen eighty three was the centennial of the Brooklyn Bridge, and the running battle between the wire inspectors and J. Lloyd Haigh, the wire manufacturer, is a classic case of construction fraud.* Several tricks were used to shift rejected wire into the approved pile. When inspectors countered these simple deceptions, wagons were diverted en route to the bridge and reloaded with the rejected material. The good wire was resubmitted for inspection. This subterfuge was finally prevented by having a man on horseback accompany each load from the inspection shed to the site. By that time John A. Roebling had died as a result of an accident on the job and his son Washington had taken his place as chief engineer. He had 150 wires added to each of the main cables in order to maintain the original factor of safety. To avoid loss of public confidence in the safety of the structure, the fraud was kept secret. Fear of bridges was not irrational at that time. In the 1870's about one of every four new bridges failed, or about 40 per year.

It often costs more to collect a sample than to analyze it.

An unscrupulous contractor can save by collecting several samples on the same visit and simply putting a different number on each one. It is not too hard to outsmart the owner who thinks he can get more than he pays for.

Laws are based on the tacit assumption that if a person can be held responsible for a structural failure, the hazard can be avoided. Is that assumption warranted? What will happen to the individual if it is possible

*For more details, see David McCullough, *The Great Bridge*, Simon and Schuster, 1972, Chapter 20.

to pin the blame for an accident on him? Will he go to jail? I have seldom heard of it happening. If he is heavily fined, can he afford to pay and continue in business? Note the prevalence of OSHA penalties of $1000 or less for serious violations; are they effective deterrents? Can owners protect their principal assets by incorporating each project? Why do so many real estate firms have the names of a dozen or more corporations on their door? Is liability insurance defeating the intent of laws that are intended to encourage responsible activity?

To illustrate some of the problems, I have chosen a few examples of occurrences within the recent past. In some cases reports are incomplete, but I am more interested in the study of the causes of accidents than in court decisions on individual cases. Complete information is seldom available; some cases are settled out of court, witnesses are not always available and when they do testify, the full facts may not be revealed. I am particularly impressed by the number of cases where someone could have averted the accident but failed to do so because he did not have direct formal responsibility and did not get involved.

A design that "looks right" may not be adequately strong, but I would give odds on the probability of an error in the design that does *not* look safe to an experienced builder. Calculations are of course essential, but sometimes intuitive judgment will indicate a weakness that would be hard to find in pages of calculations. People who do a great deal of checking tend to develop a sixth sense; they can tell that something is wrong before they can prove it.

A case in point was the collapse of the floor in a church fellowship hall in the latter part of 1981. Sixteen of the 200 people at a parish dinner were hospitalized. The floor consisted of 8 inch hollow core precast slabs, resting on an 18 inch steel beam 80 feet long, running down the center of the 50 × 80 foot floor area. The slabs were topped with 3 inches of poured concrete, and the steel beam was supported by concrete piers at each end and by three 3 inch pipe columns 10 feet high spaced about 20 feet apart.

During the dinner one column failed, shifting more load onto the other two and, when they also collapsed the 18 inch beam broke at two splices and the floor became a vee-shaped trough filled with people and furniture.

After the collapse, a structural engineer found that 6 inch columns should have been used instead of the 3 inch ones. It is surprising to me that the builder did not take the initiative to use heavier columns in the first place; the ones that failed must certainly have looked wrong to anyone with experience in construction work.

At the Hyatt Regency Hotel in Kansas City, Missouri, about 1500 people were at a tea dance. Some were in the lobby, others were on 45 ton walkways suspended over the lobby at the second, third, and fourth floor levels. Suddenly the fourth floor walkway collapsed, carrying down the one at the the second floor level that was suspended from it. The third floor walkway was independent of the others and remained intact. One hundred and twelve people were killed and 180 more were injured in the structural collapse.

The walkways consisted of concrete slabs between box beams formed of 8 inch channels on edge, welded together along the top and bottom centers of the box beams. Six rods supported the fourth floor walkway, running from the ceiling through the center welds at the top and bottom of the box beams, where they were secured by nuts and washers. Six separate rods, similarly anchored at their top and bottom ends, ran from the fourth floor walkway to the one at the second floor. When the upper one fell, the lower one went down with it.

The original design showed continuous rods from the ceiling through the fourth floor structure and terminating at the second floor walkway. The change from this arrangement to the one that failed doubled the stress at the points of anchorage of the upper tier rods to the beams at the fourth floor. As originally designed, the fastenings at the fourth floor had only to keep that walkway from sliding along the rods; the revised design put the full weight of both walkways on the fourth floor upper rod anchorages in the box beam.

Published reports indicate that the design change was suggested by the steel fabricator and was approved by the architect and the structural engineer, but the senior partner of the engineering firm stated that he did not know about the change until after the failure occurred. All concerned considered the design change too minor to require review by the city plan examiner.

After the tragedy, the debris was removed before investigators had much time to examine it, and this is perhaps understandable because there were still guests in the hotel, but the third floor walkway, which was still standing, was also removed before tests could be made on it.

Engineers were retained by a local law firm to investigate the cause of the accident, and another technical investigation was initiated by a local newspaper. The U.S. Bureau of Standards made an official study.

Dancing on the walkways may have been a factor in causing failure to occur during the tea dance, but poor structural detailing was the basic trouble. During construction, there were many competent people on the job; someone should have detected the weakness of a connection that was easily recognizable as a critical stress point.

It seems possible that an underlying problem was the complexity of the management organization; this could have led to the use of more red tape than judgment in the handling of problems such as the one that caused the collapse. Of course this is conjecture on my part, but it may be of interest to readers that the owner of record was the Crown Center Development Corporation, a subsidiary of Hallmark Cards, and that the building was being managed by the Hyatt Hotels Corporation at the time of the accident. During construction, there was a construction manager, a general contractor, and the usual subs, including the steel fabricator. Designs were prepared by a corporation formed by three architectural firms and of course there were structural engineers and the city plan reviewers in the picture. This organizational structure should not have been inherently flawed, but there were obviously enough people involved to make it difficult to keep lines of communication perfectly clear to everybody. There would be little to gain by tracing the exact chain of events that led to approval of the critical change, but the collapse of the walkways does prove the importance of having an organizational plan that brings the right questions to the right people, and this did not happen at the Hyatt Regency job.

The Kemper Arena in Kansas City, Missouri, with a seating capacity of 17,000, was the site of the 1976 Republican National Convention, and in the succeeding three years sports enthusiasts had crowded the stands many times. Then, during a stormy night when the building was empty, wind and rain combined to collapse the roof, causing over $6 million damage but no injuries.

The roof support system is distinctive. Three external trusses support a suspended assembly of seven secondary trusses that are perpendicular to the external structural members. The roof is over 300 × 400 feet, and the whole assembly was originally connected to the primary trusses by four high strength steel bolts at each of 42 points of suspension. Tests of the bolts after the collapse showed that swaying of the structure overstressed and weakened them, so that their strength at the time of failure was as little as 20% of the initial value.

In the reconstruction, greater flexibility was built into the structure, so that localized stresses would be eliminated where failure had occurred. At the same time, increasing the number of roof drains and the slope of the roof reduces the loading during heavy rain storms. Instrumentation to measure stresses during wind storms has also been added as a safety measure. Public confidence in the reconstruction work, which was done by the original contractors, was evidenced by a capacity crowd on opening night.

This failure appears to have been caused by insufficiently skilled detailing and possible stresses were not adequately anticipated. As in other examples cited, the basic concept was sound, but the detailing was deficient.

On the evening of January 17, 1978, 5000 people attended a basketball game in the three year old Coliseum in Hartford, Connecticut. A few hours later, at 4:19 a.m., the space truss roof collapsed. Miraculously, there were no injuries.

Investigations were undertaken by a committee appointed by the Hartford City Council and by an independent firm of structural engineers selected by the City Manager. My comments are based on their findings.

The engineering report included a detailed description of the space truss; it was an assembly of 960 main compression and tension members, all formed from steel angles 30 feet long. Some were in a tee form made up of two angles and the others were cruciform in section, made up of four angles. These members were used to form a top grid, 360 × 300 feet, a bottom grid 330 × 270 feet, and main diagonals connecting the upper and lower grids. A system of secondary braces, principally single angles, was intended to brace the main compression members at their midpoints. The whole truss rested on four pylons each set 45 feet in from the sides and ends.

The roof was supported above the truss on short unbraced posts of varying length to provide some roof slope to horizontal drains that led to vertical leaders within the building. A fascia system joined the roof framing to the space truss on all four sides, and it alone kept the short roof support posts vertical and resisted wind forces.

The design calculations were performed by use of a computer. The technical consultants who investigated the failure concluded that the initiating cause of the failure was design deficiency. The main compression members were all sized on the assumption that they would be braced at their midpoints, but all the exterior top grid members were either unbraced or only partially braced. Some members had loads at points that imposed bending stresses on them that were not anticipated in the design calculations. These conditions greatly reduced the ability of the compression members to resist loading without deformation and failure.

Investigators concluded that a major contributing cause of failure was a 20% underestimation of the total load on the space truss. The steel weight was underestimated, a heavier roof system was used without redesign of the truss, and other unanticipated items added to the loading. Of the 960 main members, 150 would have been overstressed by the originally estimated loading and, as a result of the actual loading, 416 were overstressed.

There were other items that could have caused failure. The horizontal rain water drains were reduced in size to 10 inches because the 12 inch pipe shown on the plans could not be readily obtained. This change and excessive deflection of the space truss that reduced the slope of the drains could have considerably retarded runoff of water from the roof during a heavy rain and added substantially to the roof loading.

In the wreckage along one of the folds where failure occurred, compression diagonal number 29-40 was found to have been interchanged with tension diagonal 29-41. Although their numbers were almost the same, the members did not have the same bolt hole spacings and consequently could not have matched the gusset plate holes when interchanged. Therefore, it is surprising that during erection the error that slipped through the shop drawing approval routine was not caught on the job. The mistake was significant. The tension member, when used in compression, had a strength of only 87 kips, whereas the compression member it replaced had a strength of 125 kips.

Detailing of the steel violated standards of the American Institute of Steel Construction (AISC Code). Investigators observed that the absence of a full-time registered structural engineer experienced with the design and construction of long-span special structures was a serious mistake and that a careful design review by either the city or an outside consultant prior to construction would have detected the deficiencies that caused the collapse. These were not new ideas; they had been rejected during the planning stage.

The independent committee appointed by the Hartford City Council observed that the contract required that the steel work be true and accurate within specified tolerances. Testimony by observers, supported by a photograph taken during construction, revealed bowing that the fabricator could have observed and reported. When the fascia was added, a change order had to be issued to permit it to be modified to fit. This misalignment should have been recognized as significant. The steel fabricator permitted only one representative to testify and he had only limited knowledge of the job-site conditions. The job superintendent at the time of erection did not testify, and certain daily reports covering the critical operations could not be located.

The committee was critical of the construction management firm, maintaining it had accepted responsibility for inspecting the truss, was inadequately staffed to do so, and had failed to perform the necessary structural inspection. The committee concluded that in failing to demand evidence that the firm was properly staffed during the construction period, the city and the construction management firm may both have "put the saving of an inspector's fee ahead of the most important factor—assuring the integrity of the structure."

No evidence that a comprehensive final inspection of the space truss had been made in compliance with the contract documents was provided by the architect or any other people in a position to do so. Testimony indicated that both the architect and the design engineer decided it was not necessary to make such an inspection.

When the city accepted the idea that the construction management firm would be responsible for inspecting the work, nobody seems to have recognized the potential conflict of interest inherent in the arrangement. If the firm had found a defect, it would have had to acknowledge responsibility for it.

The city itself had no engineers or inspectors who were qualified to check the design of a structure of this complexity, and the plan examiner felt that it would not be appropriate for him to spend the city's money to have another computerized analysis made.

I have already mentioned the importance of the owner in controlling the quality of a structure, but I want to repeat that he has the ultimate degree of control. He holds the purse strings and can do whatever he considers necessary to assure a safe structure.

J.C. Penney Company is best known as a retailer of consumer goods, but the company also has subsidiaries that write casualty insurance and a technical staff that is capable of supervising design professionals. I asked how they minimized the risk of catastrophic failures and the reply from

JCPenney

June 15, 1982

Mr. James B. Fullman, P.E.
Scofield Road & Woodley Lane
Stamford, Connecticut 06903

Subject: Safety & Code Compliance Concerning New Construction

Dear Mr. Fullman:

Your letter of May 7, 1982 has been referred to my attention. As you have pointed out one of the largest loss exposures a company can face is a catastrophic structural failure or fire. The exposure increases when a business involves serving the public within its facilities.

The JCPenney Company treats this exposure through two methods. The first method is by risk transfer (General Liability Insurance). The second method and most important can be considered loss reduction and prevention. This second method fully involves our Construction Department and Risk Management Department. Our Construction Department is located within the corporate headquarters and five regional offices. The staff includes registered architects, professional engineers, plan checkers, and full time job site personnel.

Our Department (Risk Management) and the Construction Department have developed very tight construction and occupancy specifications over the past twenty years. These specifications control methods, materials and protection within our structures.

An example of how our departments work and the application of the specifications is as follows:

1. When a new store is contemplated an outside architect is retained and given the set of owner's criteria.
2. The architect retains outside consulting engineers and together they create bid documents based on owner criteria and applicable codes.
3. The drawings, documents and actual construction are then closely reviewed by the architect, the consulting engineer, our professional field and office staff, and Factory Mutual Engineering.
4. During this process the local Municipal Building Department Personnel are also involved in the review.

As detailed above we have a considerable amount of professional people involved in a project from the time of planning right up to facility opening. With this much attention given to a project we have reduced the chance of construction or design error and received a broad review of building protection. The resultant reduction of construction problems and exposures provides an excellent return on our investment in staff.

Very truly yours,

John D. Holthusen

John D. Holthusen
Corporate Manager, Fire Protection

J. C. Penney Company, Inc., 1301 Avenue of the Americas, New York, New York 10019

Figure 6.17 J.C. Penney Company has a comprehensive plan for avoiding catastrophic structural failure or fire, as indicated by this letter from their Corporate Manager, Fire Protection.

their Corporate Manager, Fire Protection, explains a responsible policy that could well be a model for others (Figure 6.17).

6.7 Glass and Facade Failures

When glass and exterior panels fracture or fall, there is a serious danger to the public that cannot be readily controlled by sidewalk sheds or barriers close to the structure. The wind can carry shards for a considerable distance if the breakage occurs near the top of a high building.

The John Hancock Mutual Life Insurance Company's 60 story building in Boston is a classic example of glass problems. In the early 1970s, before the building was completed, shattering of glass began and all the panes were replaced in 1975. But trouble persisted; panes of the new glass continued to fall for several years. Naturally, the problem has been investigated very thoroughly, but I have seen no published report that would enable contractors and designers to eliminate similar safety hazards elsewhere.

At the state office building in Richmond, Virginia, ¼ inch thick glass spandrel panels 6×8 feet have fractured. They were made of heat-treated opaque glass that shatters into small pieces, but a hailstorm of glass fragments from a 25 story building is still dangerous, regardless of the size of the fragments.

Accidents of this kind are not entirely under the contractor's control; new methods of construction may not have been adequately developed and tested. In addition, the test of time cannot have proven them safe. For a great many years the effect of fatigue and aging of metals has been investigated. Much has been learned about how the properties of metals change over extended exposure to the elements and to stress, but comparatively little similar attention has been given to precast masonry panels or even to glass or composite panels made of materials whose individual properties are comparatively well known.

Facade failures have become sufficiently common to make the prevention of injuries to the public by falling debris a matter of concern to designers and contractors. The trouble may become worse before enough of the answers are known because it takes years for some forms of deterioration to result in failure. Meanwhile, I know of little that safety engineers can do beyond satisfying themselves that design and construction are in the hands of competent people and that work is stopped when weather conditions would make it unsafe to proceed with construction.

7

Installing Mechanical Systems and Completing the Interior

7.1 Labor Conflicts

When structural work nears completion and work begins in earnest on the mechanical systems, the number and types of workers on the site suddenly increases. This change in the composition of the work force can disrupt the safety program. Just at the time cooperative effort is most needed to avoid accidents, antagonisms can flare up to the point of involving bodily injury to workers and costly work stoppages.

The arrival of many subcontractors can spark labor-management and jurisdictional disputes. Often there are squabbles over ridiculously minor incidents. A dispute that took nearly half a year to resolve and was settled in favor of the contractor's original workers by the National Labor Relations Board flared up when union painters attempted to deliver sandblasting equipment and supplies to a job site. They were attacked, abused, and the equipment was sabotaged by members of three other unions. The managerial, legal, and union official time spent far exceeded the amount of direct labor involved on the job. It seems some unions have no sense of perspective and are completely infantile in the degree to which they concentrate on petty personal interests. Grown men should be above such action.

Problems caused by nonunion subs employed by union contractors are another source of trouble that is irrational when the sub performs a specialized service that no local union is capable of supplying. Calibrating heating or water treatment controls are typical examples. I have seen presumably mature men wasting days wrangling about who should rewire a simple control box that was originally assembled by an electrician who happened to belong to a different union than the electricians on the job. The dispute was settled by having a local electrician sit idly beside the box for long enough to take it apart and reassemble it. He was just paid to sit, not to do any work on the box.

When there is something significant to be decided, negotiation, arbitration, and litigation procedures are available. Our laws are supposed to assure fair treatment to workers and there are ample procedures for changing the laws when necessary. When inequities exist, as they always will to some extent, riots, wildcat strikes, extortion, and intimidation are not the solution. The cost of such tactics can be spent to better advantage in training programs and other constructive activity. It should be apparent by now that when industry spokesmen cannot resolve conflicts, the controversies become political footballs. Instead of solutions, we get more laws, more regulatory bodies staffed with people who write rules: that takes still more people to interpret and enforce them. I seriously doubt whether affirmative action rules, Bacon–Davis wage regulations, and even minimum wage laws have benefited more people than they have caused greater difficulty in the work place. In my opinion they are examples of the type of political action that results from failure of industry to regulate itself. At this moment, controversy over revision of the Hobbs Act, which excludes federal law enforcement agencies from labor disputes, is confusing law enforcement agencies: what should local law enforcement officers do in the face of labor violence, and what is the present status of federal regulation of such activity? Is, in fact, mayhem permitted in the name of labor unrest?

Certainly there must be better ways to resolve the tensions that pull the construction industry apart. We already have more safety regulations than can be kept in mind or be enforced. This is a contributing cause of accidents, disastrous fires, structural failures, and labor unrest that kill or maim hundreds of people every year. Until divisive forces in the industry are reduced substantially, it will be impossible to make effective progress in reducing many types of accidents and structural failures.

At the end of a labor dispute workers return to the job with different attitudes than before. The safety engineer then needs to assess possible changes in their attitudes toward the safety program. Are there animosities that must be controlled? If there were jurisdictional issues, were they resolved in a reasonable and satisfactory manner, or are the opponents likely to try to get even with each other for real or assumed inequities in the settlement? If so, what kinds of trouble may be expected?

When mechanical trades are antagonistic, the trouble may surface as interference troubles. An aggressive contractor may seize an opportunity to make part of his installation in a way that will interfere with another's work. This in turn can inspire sabotage by the other contractor, or accidents may happen to his employees. His equipment may fail to operate when needed, raising the question of whether it was defective or deliberately damaged. There may be minor troubles; if a scaffolding plank breaks, was it cracked by a deliberate overload and left as a booby trap? Who left a tool on the top of the step ladder? Was it an accident when a tool fell and struck someone? Was it a manufacturing defect that caused a grinding wheel to fragment at normal speed? Whose illegible signature is on the receiving slip for missing parts needed to complete a sub's contract? Some people have disagreements, settle them, and forget about it. Others bear grudges long after the immediate project is finished, so that trouble on one job may carry over onto the next.

Safety engineers can't work in isolation, writing slogans, alone in an ivory tower. Nothing beats a good grapevine, a source of information on what is really going on. I do not mean a spy system, or informers; just the normal rapport that exists between people who trust one another and want to avoid trouble on the job. If "A" and "B" are known to have it in for one another, it is sometimes possible to transfer one or both to jobs where they are out of contact with each other. Sometimes it is sufficient for workers to realize that supervisors are aware of what is going on to keep them from trying to stage "accidents."

As long as there are workers on the job, there will be problems that plague foremen and safety engineers alike. Maintaining a reasonably cooperative attitude between contractors and between individual workers is part of the safety program. In the comments on additional safety matters, I will assume that peace, (if not unbroken tranquility) has at least been achieved.

7.2 Positioning Parts of Systems

When major components of large systems must be handled, their weight and bulk make it obvious that suitable rigging methods must be used and that the work must be properly planned. The weight of large pipe, valves, and fittings also precludes manual handling. Consequently, these operations involve safety hazards but accidents are comparatively rare. The probability of an injury is greatest when the item to be handled is heavy or bulky, yet workers or their foreman mistakenly decide they can manage without more experienced help. Then, when suitable rigging equipment is not used, there are back injuries, crushed fingers or toes, and damage to the equipment being handled. When the load weighs in the 200 to 2000 pound range, the temptation to use makeshift rigging methods seems greatest and while simple equipment such as chain hoists, rollers, dollies, and jacks may be entirely suitable, not everyone knows how to use these tools safely. It is a joy to watch professional riggers at work; they seldom seem to be working very hard, yet every move counts, and the item being moved responds exactly as they know it will. Their know-how is worth the extra cost when it is apparent the job will be difficult for the installers to tackle alone. Injuries are expensive and, even if cost is the only consideration, it does not pay to risk an accident.

Accident prevention requires imagination; it is better to anticipate many troubles that never materialize than to have one crop up that was not expected. When a major component of a mechanical system is to be delivered, be sure someone has checked the route for obstructions. Highway engineers are finding that literally hundreds of bridges are in need of major repairs, and just because a bridge has not been posted to restrict the weight of vehicles does not mean it is safe. A survey of the bridge's condition may be underway, or the bridge may be in questionable condition. A call to the highway department or public works agency could avoid serious trouble. I have listed some crane hazards elsewhere; see Section 6.2.2. Could the proposed work be done safely if the weather took a turn for the worse? Should plans be made for rescheduling the job, or should foul weather plans be made?

I suggest that the plans be checked to see whether there was ever any change in the position assigned to heavy equipment and, if so, whether the change was approved by the structural engineer. At the Hartford, Connecticut Coliseum where the space frame roof collapsed, the cables supporting a five ton scoreboard were wrapped around structural members apparently without any engineering consideration having been given to their proper attachment. I could not determine from investigation reports whether the load had even been considered in the structural design.

On a recent job, wide load permits were required to move equipment on city streets and when, in the early morning hours, the work was interrupted by the police, the workers had no permits. The permits were in the office and, by the time they were found, rush hour traffic had become badly snarled and the work was delayed. All the big problems had been solved; the nit-picking ones caused the trouble.

7.3 Small Tool Safety

A generation or two ago, if employers had safety programs at all, they were seldom of much value, particularly if any appreciable cost was involved. It was largely up to the employee to look out for himself. More often than today, fathers expected their sons to follow in their footsteps, and taught them survival skills. My father, who was a toolmaker before becoming an engineer, took pains to point out the dangers inherent in the misuse of every tool he showed me how to use. Later, as an apprentice in a machine shop, I was fair game for pranks, but the old timers still took pains to keep me out of real trouble. They explained how they had lost fingers, gotten scars, or how friends had lost their lives on the job.

Attitudes are shaped by such early training; manual skills are made obsolete by technological developments, but habits of thought remain. Out of all the instruction a person receives over the years, he remembers only a small part, but his basis for selecting what to remember is influenced by people whose judgment he values. When the need for considering safety as an important work skill is impressed on a trainee, he will probably continue to be receptive to safety instruction. But, if his first instructors teach him unsafe work habits, it is doubly hard to change them later. In recent years we have certainly had more employer involvement in safety, but I think the personal worker-to-worker relationship has often deteriorated, and training that was formerly given by a father is often obtained at a trade school or government-sponsored job training program.

Schools probably shorten the training period, and no doubt students are introduced to more tools today than were even in existence 20 or 30 years ago, but I doubt that equivalent progress has been made in teaching survival skills. Let us not forget that on the job it is the ability to survive and to continue to learn that are most important in the long run.

I think some instructors expect students to recognize safety hazards by using their common sense when they have no background that would enable them to do so. A farm boy, raised around horses, does not have to be told the difference between a stallion and a gelding. A youngster who works as a helper on residential construction jobs picks up more incidental information than he realizes. But when a young man or woman goes to a trade school and expects to become a carpenter, electrician, or other craftsman, and has no previous contact with the construction industry, the first job is in a strange world in which he has little ability to recognize dangers that would be apparent to anyone who grew up in that environment. He has to be taught basics.

I recently visited two school shops, both quite well provided with modern tools, but with many unguarded vee belt drives and other hazards that would not have been tolerated in a well-run industrial plant. Any OSHA compliance officer would have given a manufacturer with such conditions numerous citations. When I mentioned this to the instructors, they shrugged their shoulders and said nobody had gotten hurt yet. That seemed to them to be adequate evidence that there were no danger spots. I doubt whether such instructors can be expected to encourage students to develop safe work habits. Possibly they were the exception, but a study of 2528 accidents in the construction industry in Maryland in 1976 revealed that two thirds of the injured workers had been on the job one month or less and that 90% of the accidents happened to people who had been on the job six months or less. Safety engineers and foremen need to be alert to the needs of the new employee, even when the worker has had experience elsewhere.

Still, when an attempt is made to explain a safety hazard to a new employee, he is apt to resent it. Both young and older workers often seem to think they know it all, and it is practically impossible to get them to listen. They refuse to accept the fact that many injuries result from the improper use of even the simplest of tools.

Consider, for example, the ordinary flat blade screw driver, probably the most abused of tools. A true craftsman insists that the blade fit the slot of the screw and that the tip be square to the shank. He buys tools made of good steel and has enough sizes to handle the work he expects to do. When he needs a pry bar or cold chisel, he uses one and saves his screwdrivers for their intended use. But how many workers who call themselves mechanics have screwdrivers that are sharpened more like a chisel than a square-edged tool? How many hammer on the handles, or have bent the shanks by prying something with them? Such tools often slip and leave a sharp sliver of metal on the head of the screw where it can become embedded in the first thumb that passes over it. Such workers are often seen tightening a screw while holding the work in one hand, forgetting that if the screwdriver slips it may pass entirely through his hand. Such stab wounds are painful, disabling, and apt to become infected.

Improper use of wrenches is another common cause of accidents. Many workers use a pipe wrench to tighten bolts and nuts; they may damage either the fastener or the tool. Then, when the tool slips, knuckles may be skinned or more serious injury may occur. Trainees may have so little knowledge of wrenches that they do not even know the names of the different types, let alone their proper use. They need to be taught how to anticipate when a wrench may slip, and how to avoid injury when it happens. The principal problem of the safety engineer is to get those who are most in need of training to pay attention. Those who have already suffered a painful injury are more apt to listen than those who have been lucky thus far.

Skil Corporation, a subsidiary of Emerson Electric Company, introduced the first portable electric saw in 1923 and on the basis of their long experience in the field, they recommend the following specific precautions for users of their circular saws:*

Disconnect plug from power supply before changing blades, making cutting depth or cutting angle adjustments, inspecting, cleaning, or when saw is not being used.

Do not use dull or damaged blades.

Keep hands away from cutting area.

Keep blade guards in place and in working order.

Never clamp or tie the lower guard into the open position.

Raise the lower guard only with the Retracting Handle or Safety Guard. Lift Lever to make sure it moves freely and does not touch the blade. Do not operate saw if lower guard does not function properly.

Do not run the saw while carrying it at your side.

Secure wood while sawing; never hold pieces for cutting in your hand or across your legs.

*Skil Operating Guides, courtesy Skil Corporation, 4801 West Pederson Avenue, Chicago, Illinois 60646.

Keep second hand on motor housing or auxiliary handle, not near blade, when guiding saw with two hands.

Make certain depth and bevel adjusting wing nuts are tight and secure before making cut.

Position the electric cord away from line of cut, both above and below.

Do not operate saw with a damaged cord.

(See Appendix E for recommended gages of extension cords.)

In addition to the foregoing, Skil Corporation recommends that users of their tools observe a number of general safety precautions, among which are the following:

Use safety glasses, and face or dust mask if cutting operation is dusty.

Never carry tool by its cord or yank the cord to disconnect it from receptacle. Keep cords from heat, oil, and sharp edges.

Secure work; use clamps or vise to hold work. It's safer than using your hand and it frees both hands to operate tool.

Don't overreach; keep proper footing and balance at all times.

Remove adjusting keys and wrenches; form habit of checking to see that keys and adjusting wrenches are removed before turning tool on.

Avoid accidental starting; don't carry plugged-in tool with finger on switch. Be sure switch is OFF when plugged in.

Avoid gaseous areas; do not operate portable electric tools in gaseous or explosive atmospheres. Motors in these tools normally spark, and the sparks might ignite fumes.

When tool is in use outdoors, use only extension cords suitable for use outdoors and so marked.

When abrasive wheels are used to cut tile, marble, or blocks, follow the manufacturer's instructions carefully to avoid damage to the blade. If abused, it may fragment and the broken pieces may be thrown for a considerable distance. The operation may also create harmful dust, and appropriate measures should be taken to keep workers from inhaling it. Lung damage can result from failure to observe this precaution.

Reciprocating power saws, using either jigsaw or special bayonette blades are difficult or impossible to guard. Manufacturers generally recommend that the blade be moving at full speed before a cut is begun, and that particular care be taken to avoid binding of the blade in the work. This is most apt to occur when cutting around a curve or when approaching the end of a cut; always support the scrap or free part of the material being cut. Wear goggles and try to stay clear of the path that might be taken by parts of the blade should it break.

Only workers who are completely familiar with the manufacturer's safety recommendations should be allowed to use radial arm or table saws. When ripping stock, at the end of the cut the scrap can be propelled back toward the operator if the antikickback pawls do not work. Always anticipate this possibility and stay clear. Keep others out of the danger area also; I have seen scrap fly like an arrow even when the tool was being operated by a factory-trained demonstrator.

Workers often suffer minor injuries when starting a cut with a hand saw. If a worker does not know how to do it without risk of losing the end of his thumb, he may never have been properly instructed. Some graduates of training courses do not even know how to tell which end of a hack saw blade should be nearest the handle, and their knowledge of the different types of blades is equally astonishing by reason of its absence. When such ignorance of basics is observed, it is certainly to be expected the same worker will be equally ignorant of safety precautions that should be second nature to any mechanic.

Some motors have an automatic overload cut-out switch that restarts the motor again as soon as it cools off. I strongly recommend that all such motors be removed from power tools and replaced with overload protection that requires manual resetting.

There is some disagreement among mechanics about the safest way to use wood chisels when driving them with a mallet, and similar disagreement applies to the way to do chipping work with a hammer and cold chisel. Some prefer to cup the chisel in their hand, knuckles down, and with the thumb pointing toward the cutting edge. If they miss the head of the tool, the mallet or hammer strikes the muscular part of the hand at the base of the thumb. Others feel that they have better control of the tool if they wrap their fingers and thumb around it, even though they realize that if they miss hitting the chisel they will bang their knuckles. I can see merit to the first method if the worker does not feel confident that he can hit the chisel without looking at it, but gaining that confidence is somewhat akin to learning to use a spoon without being able to see where your mouth is. Once the knack is acquired, it no longer becomes necessary to watch the head of the chisel, and attention can be concentrated on the cutting edge. I think the second method provides better control of the tool. But, to each his own way.

On some jobs I have had difficulty finding a worker who knows how to handle a sledge hammer while someone holds a star drill or chisel. When they miss the chisel head or strike it a glancing blow, someone is apt to get hurt even when the chisel is held by tongs or heavy pliers. The same problem arises when some paving is to be broken and no pneumatic tool is available. A strong young man can break the handles on sledge hammers faster than a good carpenter can replace them. They seem to want to drive the head of the hammer right through the work, and their aim is often so poor that they are a hazard to everyone in the vicinity. The only suggestion I can make is to give the job to someone else, even if he is twice the age of the first worker. He may take many more rest breaks, but will probably get the work completed faster. I doubt the feasibility of teaching anyone to use a sledge hammer; modern impact tools have almost entirely replaced the sledge and the safest procedure is to recognize that fact.

Having made some uncomplementary remarks about poorly trained youngsters, it is only fair to point out that old timers tend to become attached to old familiar tools and to use them when they are unsafe. A good hammer, hatchet, or saw deserves a new handle if the old one is split. A cold chisel with a mushroomed head should be reshaped or replaced before someone is struck by a fragment of steel. Wrenches that have been bent out of shape and screwdrivers that have passed their retirement age are still to be found in many tool boxes. Even the best mechanics sometimes need the spur of a tool box inspection to convince them that they need to replace some unsafe tools. Similarly, workers will readily agree that frayed cords should be replaced, but they may disagree about whether they or the company should pay the bill. Remember that it is the employer who is responsible for maintaining safe working conditions, and be guided accordingly.

Workers whose experience with electric drills has been confined to the ¼ or ⅜ inch size may be unaware of the dangers associated with the use of slow speed high torque drills made for handling larger diameter bits. When a drill bit or hole saw catches in a piece of sheet metal that has not been clamped in position, the work piece may swing around and inflict a serious wound. On the other hand, if the work is immovable and the tool catches, the worker can be thrown off a ladder or may suffer a wrist or arm injury. He may also drop the tool on someone working below. Every worker must expect that the bit will catch in the work from time to time, and he should be prepared for it to happen. If he is working where a sudden twist of the tool might cause him to fall, he should wear a safety belt. Sometimes the work should be moved to a spot where the necessary job can be done safely. Accidents often happen when someone tries to save a few minutes and takes an unnecessary risk of injury.

Sheet metal workers often use a nibbler to cut openings in duct work and these tools are not apt to cause injury to the user, provided he avoids exposing his eyes to the sharp crescent-shaped slivers of metal that are discharged by the tool. However, those bits of metal can be dangerous to others and can easily damage open electrical equipment.

The bits used in pneumatic tools should not only be securely fastened in place, but workers should avoid pressing the trigger until the tool is in contact with the work. Even when the worker recognizes that he needs goggles and a respirator, it is generally the employer who must take the initiative to get him to wear them.

When prolonged use of chipping or grinding is done, particularly in cold weather, vibration white finger disease is an occupational hazard. It can affect nerves, bones, muscles, and arteries. It is not confined to the hands or fingers. In severe cases the skin may become ulcerated and gangrene may set in. During early stages of the disease, workers may fear loss of their jobs and fail to report distress. The largest tools may not necessarily be the most likely to cause injury; the ratio of tool weight to the force of the blows, the number of blows per minute, and the duration of exposure per day all affect the severity of stress on the worker. The U.S. Department of Health and Human Services, National Institute for Occupational Safety and Health (NIOSH) in Cincinnati, Ohio, 45226 has published valuable reports on this disease.

Users of pneumatic tools should wear hearing protection in most instances (see Chapter 3, Section 3.3.3).

Tools that use explosive charges to drive special fasteners into steel or concrete have been used extensively for more than 25 years, and their design has been highly refined to minimize safety hazards. Still, it must be remembered that any tool capable of driving special steel fastenings through structural steel over half an inch thick has to be handled properly. Powder-actuated tools must always be used under conditions that will prevent the fastener from becoming a projectile or dislodging masonry that could strike anyone.

A teen-age girl was working in a fast food restaurant when a pin came through the wall; it had been driven by a worker using a powder-actuated tool in an adjoining building. The pin struck her in the temple and she was in critical condition for some weeks. It was only after several months in the hospital that she was able to walk again. The tool should not, of course, have been used on a wall that could not be confidently relied upon to prevent the fastener from penetrating it completely and becoming a projectile.

Accidents have also happened when the tool was not held perpendicular to the work surface, so the stud was deflected and ricocheted off the metal or masonry. If a pin is too slender for the application, it may curl as it penetrates and spall off adjacent concrete. Use only fasteners approved by the tool manufacturer; never use ordinary nails, or even the hardened steel ones sold for driving into masonry.

Fastenings to masonry can be safely and securely driven when conditions are right, but care must be taken to avoid common dangers. A properly set pin penetrates no more than one third the thickness of the solid masonry. Consequently, they should not be driven into hollow blocks or vertical mortar joints; they may fly completely through the wall. If used at all, pins should be driven through a steel clip, channel, or similar member that will restrain it if the masonry is too weak to do so. If it is necessary to set studs into masonry joints, drive them into the horizontal joints only, and be sure no workers are on the far side of the wall while the work is being done.

When driving pins or studs into solid masonry, avoid striking reenforcing bars or anything that could bend or deflect the fastener and cause it or spalled concrete to fly through the air. Do not drive pins close to corners; they may split off and allow pieces of masonry and the fastener to injure someone. Always follow the manufacturer's recommendations for the use of shields and for the selection of stud sizes and charges. Some masonry materials are too hard and brittle to accept pins; others are too soft to retain them properly. Testing procedures are included in the training given workers by manufacturer's representatives.

In contrast to the requirement that pins penetrate no more than one third of the way through concrete, studs or pins should penetrate completely through steel members if maximum holding power is to be obtained. There is danger of a ricochet if the fastener is driven too close to an edge or drilled hole.

These and many additional instructions will be given to workers by tool manufacturers and, before anyone is allowed to use a powder-actuated tool, his supervisor should be sure he has a Qualified Operator's Card, signed by an authorized instructor and identifying the worker and also the specific make and model he has been trained to use.

Safety engineers should watch workers to be sure they habitually follow commonsense safety procedures. A loaded tool should never be left unattended or carried from job to job; the trigger should never be touched until the tool is in position to be fired; every tool should be handled as though it were loaded unless it is apparent that it is not. Horseplay should, of course, be prohibited. Even if the tool is empty, pointing it at a person may make him feel threatened, and his instinctive reaction may be to defend himself with whatever tool is handy, or with a quick punch, kick, or knee move that is completed before the threatened person consciously wills it. Goggles should be worn by workers who might be exposed to flying

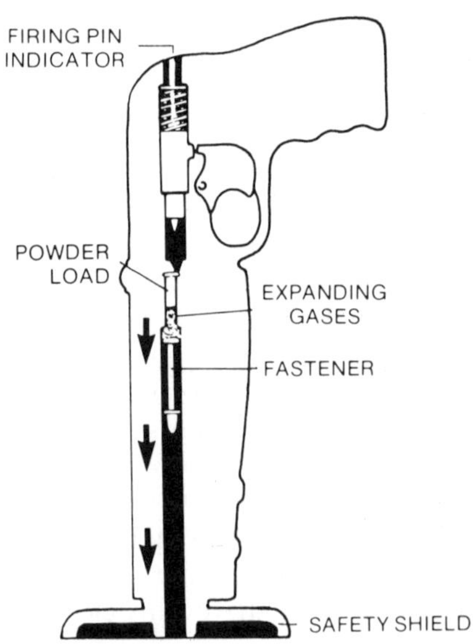

Figure 7.1 A powder-actuated tool working on a high velocity direct-acting principle. This design provides the most efficient method of fastening. Drive pins and threaded studs are driven directly into concrete or steel by the force from a special industrial powder load. Fastenings can be driven 1¾ inches into concrete or through steel ¾ inch thick. Always follow manufacturer's recommendations regarding safety, training of workers, and selection of equipment. (Drawing courtesy Ramset® Division, Olin Corporation, East Alton, Illinois 62024.)

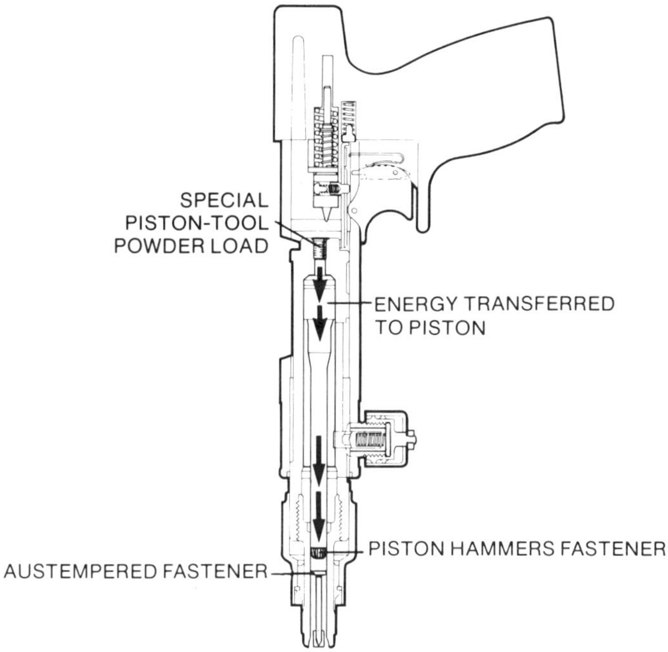

Figure 7.2 A powder-actuated tool working on low velocity indirect-acting principle. The model shown is a light duty, low velocity piston tool that utilizes the energy from a special piston tool powder load to drive a piston that in turn sets a fastener with a single hammer blow into steel or concrete. Always follow manufacturer's recommendations regarding safety, training of workers, and selection of equipment. (Drawing courtesy Ramset® Division, Olin Corporation, East Alton, Illinois 62024.)

chips, even though they themselves are not operating the fastening tools.

Those in charge of safety matters should also be sure the tools issued to workers are appropriate for the job to be done. For example, the most powerful tools use the explosive charge to drive the fastener down the barrel at high velocity, whereas tools for use where less power is needed usually have a piston in the barrel and, when the cartridge is fired, the piston strikes the fastener, driving it home (Figures 7.1 and 7.2). Every tool can be used over a certain capacity range by using different charges or making adjustments to the tool itself, but such modifications can be made only within limits imposed by the design of the tool. Even a worker who is trained in the use of a tool may assume that when it is issued to him it is suitable for the job. He will presumably make a test, using the lowest power charge before working up to the one that is most suitable for the job, but if he has been issued the wrong tool, even the initial test firing may be dangerous.

Someone has to be responsible for the security of the explosives; they are typically too powerful for use in ordinary handguns and anyone who steals a box of cartridges may be unaware of the danger to himself or others if he fires them. Leaving boxes of cartridges where they may tempt other workers to appropriate them creates an attractive nuisance.

Subpart I of 29CFR 1926 deals with hand and power tools at construction jobs; paragraphs 1926.300 through 1926.354 are applicable. In addition, OSHA has identified portions of the general industry regulations, 29CFR 1910 as applicable to construction. Paragraph 1910.212 applies to all machines, paragraphs 1910.301 and 1910.302 apply to hand tools, and 1910.244 supplements the rules in 29CFR 1926.304 on jacks. The regulations cited also require compliance with the American National Standards Institute (ANSI) Safety Code for Mechanical Power Transmission Apparatus, B15.1-1953 (R1958), ANSI A10.3-1970 on Safety Requirements for Explosive-actuated Fastening Tools, and ANSI B7.1-1970 Safety Code for the Use, Care, and Protection of Abrasive Wheels. I recommend that safety engineers familiarize themselves with the foregoing; they will bring to mind many familiar hazards that might easily be forgotten. One rule that is very often cited by OSHA compliance officers requires that bench-mounted grinders have rigid work rests and that they be no more than ⅛ inch from the wheel. Frequent adjustment is of course necessary to avoid deviation from this regulation, and it is the basis for many accusations that OSHA officers are more interested in nit-picking citations than more serious offenses.

7.4 Cutting and Patching

Safety engineers sometimes need a dozen pairs of eyes to catch all the safety hazards caused by cutting and patching work. Watch particularly for openings through which fire might spread from one room to another, or through which smoke could pass. Subcontractors may consider fire stops just useless obstructions to their work and fail to do any patching work. They may have equal disregard for structural members. If bar joists are in the way of duct work or piping, some workers will cut some bars and bend them out of the way, with total disregard for the effect of such activity on the safety of the building. They "prove" to themselves that this is permissible because nothing is apt to happen immediately when one bar is cut; so if it can be done to one bar, why not to a bar in every joist all across a room?

If sleeves were forgotten before concrete slabs were poured, it may be necessary to cut a hole where a reenforcing bar is in the way. Whether or not this is permissible is a decision that should not be left to the judgment of the subcontractor who needs the additional opening.

When work that has been completed by one subcontractor is in another's way and must be temporarily removed, the contractor who got there first may think he has an opportunity to profit by overcharging for removal and replacement work. If this is a possibility, the second contractor may watch for a chance to do the work himself, giving little thought to how well he does it. Pipe hangars may be omitted or connections may be improperly made.

In case of fire, unsealed floor openings can also increase water damage.

7.5 Housekeeping

In Section 4.6 I mentioned the importance of early enforcement of agreements on trash removal and good housekeeping on the site. This topic must be reviewed often, particularly when new subs become involved.

At a large job in Seattle, there was such poor housekeeping and so many other safety hazards that the electrical workers and iron workers walked off the job in protest. The general contractor admitted that conditions had gotten so bad that they could not expect the subs to do the cleanup and the general contractor had to take over.

"Housekeeping" is a word that tends to turn people off; workers are apt to think of it as demeaning, and housewives prefer to be called homemakers. In a hotel the housekeeper is a supervisor, not a person who does any of the actual cleaning. Little wonder that construction workers seldom clean up their trash unless required to do so. But accidents in search of places to happen seek out rubbish like vultures looking for carrion. Loose trash is a tripping hazard, broken bottles, boards with protruding nails, and scraps of pipe are too obviously dangerous to require discussion.

Rubbish also takes up valuable space and encourages petty theft. Small tools and personal belongings of workers "accidentally" get into trash piles until they can be "found" by the thief. It is also easier to determine who is responsible for cleaning up if it is done every shift before one crew can put trash on a pile left by others. It is hard enough to get people to clean up their own refuse, but it is practically impossible to get them to dispose of waste left by others, even when they all work for the same contractor.

OSHA regulations require that the workplace be kept clean. See 29CFR 1926.250 and 1910.176 for typical rules. They do not say how to clean up or when; they merely state that it shall be done. Note the use of the word "shall," rather than "should."

Time has to be allocated to cleanup work, and I suggest that it be done at the end of every day. If it is done often enough, it takes very little time and, if every worker realizes that he will have to help, it discourages needless disorder. Also, it is easier to get started the next morning if the work place is in good order.

Workers can be expected to neglect this cleanup work to see if they can get away with it. If nothing happens, there will be a double accumulation the next day, and so on. If they can dodge the job long enough, they may win the battle; someone will have to be hired to do the job.

When mechanical systems are being installed, housekeeping is apt to become a difficult problem because there are so many trades doing work in the same general areas. It is hard to tell whose rubbish is lying about, and it is even harder to keep the area cleaned up when everybody's trash is mixed together on the floor. I have no pat answer to this problem; jobs differ. When there is a lot of rubbish resulting from one sub's work, such as crates, pallets, and scrap materials, he may be willing to clean up the area in return for some equivalent favor from others working in the same vicinity, such as the use of a rolling scaffold or special tool. Possibly it can be handled by the general contractor's accepting responsibility for all rubbish removal. Occasionally the cleanup job can be done area-by-area, with one sub cleaning up everything in his assigned space. The cost of some cleaning help can also be shared by all subs if they can agree on how to split up the cost.

Regardless of how it is accomplished, trash removal and orderly maintenance of the work area is a safety requirement. If it is neglected, accidents are almost certain to happen.

Food containers and wasted food can attract rodents and insects, both of which are likely to cause mechanical problems as well as health hazards. When piping systems are being installed, there are innumerable opportunities for infestation and also when equipment is put into service, automatic control valves fail to work, strainers get clogged, or potable water lines cannot be sterilized to the satisfaction of health authorities.

As the job nears completion, the kind of rubbish that is generated tends to be more dangerous than at early stages of construction. When masons are at work, they leave mortar, metal lath, and corner bead, along with some small amount of combustible material, but when electricians are at work, they are producing scrap ends of conduit that can roll and cause a person to fall if he steps on it. They will be discarding cartons in which fixtures and system parts are shipped, all of which are flammable. Plumbers leave similar rubbish, some of which may be saturated with cutting oil. Insulation may include glass fibers that are irritating to the skin, and there may be some sharp fasteners with large washers on which a worker may accidentally hurt his foot.

In machine rooms, as the floor space becomes more crowded with equipment, there are more nooks and crannies in which rubbish can collect; it is no longer easy to make a clean wall-to-wall sweepup of the space.

Before partitions are erected, it is relatively easy to illuminate the work area, but the nearer the job is to completion the harder it becomes to avoid having dark spots where a worker cannot see tripping hazards. Stairs are often a danger area, particularly until the treads are finished.

Inspection also becomes more difficult; it is easy to see whether a large open area has been cleaned up, but it takes time to check a large number of individual rooms. By the time the last one has been checked, the first may be filled up with trash collected from others.

7.6 *Security*

There is often a relationship between security and safety. Sometimes it may be obvious. If a worker's ladder is stolen, he is likely to take chances by improvising a way to get his job done. At other times, disappearance of materials or the personal property of workers may be an indication that someone is a drug addict and a dangerous person to have on the site. Petty theft in such cases may not immediately be recognized as a warning that safety could be affected, but there is certainly that possibility.

While mechanical systems are being installed, many small items that are readily stolen will be found on every job. Some materials are useful for home repairs, so small pipe fittings, valves, pop rivets, and the like may be stolen. Other items, particularly those made of copper or brass, have scrap value, and a light-fingered worker may carry off valuable parts for which he gets only a fraction of their cost from the local scrap dealer. Frequently, it is not the actual value of the stolen items that causes the most trouble. If ten valves are needed and only eight can be found, the time lost in completing the job, and the paper work involved in getting the replacements may far exceed the original cost of the items.

Good housekeeping and inventory control will minimize these difficulties. In some inner cities, petty thievery can be a major problem; anything that is not nailed down will disappear.

Contractors should work out reliable methods for receiving shipments of small parts. Many losses have occurred because suppliers were instructed to make deliveries directly to the job site. It is unrealistic to expect delivery men to be very much concerned about who signs for a shipment. If they have a delivery slip that is to be signed by a certain person, anyone who answers to that name may scrawl something on the "received by" line, take the box, and do with it as he likes. The delivery man seldom knows the person to whom he is to make the delivery, nor is he likely to remember the person who signed the slip.

People who are authorized to sign for shipments should check the number of packages carefully. Cartons are often banded together; if the list shows one package, the delivery man may separate two cartons that are banded together, tender the one with the label on it to the receiving clerk, and leave the other one in the truck. If the driver is overly anxious to get a signature, watch out. He may be illegally parked and have a good excuse for hurrying, but then again he may just want to avoid having the shipment checked. Watch for obvious weight errors.

If security is a problem, I suggest that contractors have small items delivered to their office, away from the job, even though someone has to bring them to the site later when workers are ready to install them permanently.

When it is necessary to give subcontractors keys to locked rooms, do not give them master keys, even though the person to whom the key is given is perfectly trustworthy. It may be stolen from him and duplicated, or he may lose it. It is better to go to the extra trouble of keeping track of many keys than to have to rekey all the cylinders if a master key goes astray. I suggest that local locksmiths be asked to cooperate in controlling duplication of keys. People who want to have a key to which they are not entitled may be expected to patronize a store that advertises that keys will be made while you wait. If the blank used in locks at a new job is chosen so it will be easy to identify the keys to its locks, some arrangement may be made with them to be "temporarily out of the right blank" until he can confirm that the order is legitimate. If several master key zones are appropriate, using different blanks or different makes of locks may be desirable so that if one master key is lost the extent to which security is endangered is lessened. Requiring people to sign for every key and account for it when asked to return it is just common sense, but is often neglected or handled in a haphazard manner. In some localities it may be possible to leave all doors unlocked and issue no keys to anyone until the building is turned over to the owner. This used to be possible in cities where nothing would be safe today, and I have therefore become perhaps overly cautious. One tends to remember the things that disappeared and to forget about all the unlocked items that were never disturbed.

Professional security organizations may be needed in some cases; the suggestions I have made are more applicable to the jobs where theft is a minor but annoying problem. Even when uniformed guards, checkout systems, night watchmen, or guard dogs are used, there is no completely reliable system; if workers want to steal, there is no way to keep them completely honest. If policing is overdone, it may actually make it a contest between workers and guards and increase the incidence of theft. Some people respond better to expressions of trust than to threats of arrest. There is no complete answer; just keep an eye on the situation and don't let it get out of hand. Once workers get used to stealing it is hard to stop them.

It goes without saying that workers with money in their pockets attract people who do not care how they can take it away from them. Some workers feel that once they have earned the money, they are entitled to do with it as they choose but, if they spend it for drugs and return to work, their freedom to do as they like is no longer their concern alone. By attracting drug dealers to the area and endangering others on the job, they have gone beyond their legitimate rights. Furthermore, illegal activity does not necessarily stop at the fence around the site. It is a small step from dealing on the street to dealing on the job; dope peddlers aren't fussy.

I have delayed commenting on these problems because they are more likely to become most difficult to handle as the job reaches the final stages of construction, when there are more workers, more paymasters, and more enclosed spots on the job where people can be out of a supervisor's sight.

Some negotiation with workers may be necessary to gain their acceptance of the idea, but I think one key to minimizing crime problems is to reduce the amount of cash in circulation at any one time. Direct deposit of paychecks to branch banks near worker's homes can let them get cash as they need it without getting it all at one time. More banks than ever are staying open one or more evenings a week, and automated tellers that can be used to withdraw money at any hour of the day or night are quite common. The old complaints about difficulties of cashing checks will still persist, but they are no longer as legitimate as they were in days when workers could go only to check cashing services and pay a fee to get their money. To a large extent, an educational problem exists. Many workers still have little understanding of how banks function, and I can readily appreciate their confusion when banks try to outdo one another in devising new names for their variations of checking, savings, and combined checking-savings accounts with and without automatic loan privileges, time deposits, and all the rest. I do not think there is any one best way to handle payrolls; in fact, it is possible that having several systems in operation at the same time may be better than any one plan. By diversifying the payroll so that muggers cannot so readily predict who will be carrying much cash, the crime problem may be reduced. Employers who are concerned about the safety of their employees can do a great deal to help by simply explaining how workers can use the banking system to their advantage.

7.7 *Confined or Enclosed Space Hazards*

The National Institute for Occupational Safety and Health (NIOSH) has released a 68-page "Criteria for a Recommended Standard for Working in Confined Spaces," DHEW (NIOSH) publication number 80-106. I recommend it to safety supervisors.

In OSHA regulations 29CFR 1926.21(b)(6)(ii), a confined or enclosed space is defined as one "having a limited means of egress, which is subject to the accumulation of toxic or flammable contaminants, or has an oxygen deficient atmosphere. Confined or enclosed spaces include, but are not limited to, storage tanks, process vessels, bins, boilers, ventilation or exhaust ducts, sewers, underground utility vaults, tunnels, pipelines, and open top spaces more than 4 feet in depth such as pits, tubs, vaults, and vessels."

OSHA requires that employees be instructed regarding hazards, precautions, and use of protective and emergency equipment. Workers must also comply with any specific regulations that apply to work in dangerous or potentially dangerous areas.

Working in a confined or enclosed space is inherently dangerous because human beings can easily lose their ability to react normally when exposed to even seemingly minor variations from typical working conditions.

When a worker is in a confined space, he needs to be alert and well coordinated. Anything that interferes with his physical well-being can cause an accident that would not otherwise occur.

Air quality is just one consideration. Humans are accustomed to living in an atmosphere that contains 20.9% oxygen, 78.1% nitrogen, 1% argon, and only minor amounts of other gases. If the oxygen content of the air falls to 17%, breathing becomes noticably more difficult and heart action accelerates; 16.5% is often considered the minimum to sustain life. If the oxygen level falls to the 14–16% range, muscular coordination becomes poor, workers tire very rapidly, and breathing becomes irregular. At still lower oxygen concentrations, workers often become nauseous, have convulsions, lose consciousness, and may die very quickly. Even as little as a four minute exposure to severely restricted oxygen can cause death or permanent brain damage. The minimum allowable oxygen content of air permitted by OSHA in tunnels and shafts is 20%. See 29CFR 1926.800(c).

Artificially increasing the oxygen content of the air can be disturbing as well as dangerous. Hair and clothing can absorb oxygen and suddenly begin to burn violently. The explosion hazard that results from contact with oil and oxygen is of course well known to welders.

There can be significant changes in the oxygen content of the air in even apparently safe locations, such as open-top tanks or shallow pits. Gases that are heavier than air can pocket in such spots and many such gases are odorless. If a worker becomes disoriented and faint, he can injure himself. An unconscious person can drown in only a few inches of water.

Problems associated with oxygen deficiency become more pronounced at high altitudes. Contractors who contemplate doing work in mountainous areas should get medical guidance in setting up safety procedures.

While in an enclosed space, a worker may be in an awkward position, be encumbered with safety gear, or exposed to unusually hot, cold, or humid conditions. Overexertion can result in heart failure, heat exhaustion, or muscle cramps.

Unanswered questions often remain after an enclosed space accident because the people involved do not survive. Early in 1981 a contractor on Long Island was engaged in epoxy coating the inside of a 500,000 gallon capacity water tower, 150 feet high, with a 4 foot diameter open pipe from its bottom to ground level. Two men were working inside the tank, using rope-supported seats, and their supervisor, one of the members of the firm, was on the outside of the tank observing the work. There was a ventilation fan at ground level, forcing air into the vertical pipe and until 5 p.m. work seemed to be progressing normally. Then one painter fell to the ground through the drop pipe at the center of the hemispherical bottom of the tank. The other painter also fell, but escaped the open pipe. When the accident occurred, the foreman climbed down an external ladder, went to a neighboring house, and called for help. He then climbed back up the ladder and entered the tank, where he remained until the local fire department rescued him and the surviving painter about 11 p.m.

He died the next day of exposure and exhaustion. Why the open drop pipe was not covered by a grating or otherwise guarded has not been announced. The foreman was out of touch with possible rescue workers, and was unable to get his men out of the tank by himself. The ventilation system was not working when the fatalities were investigated. Planning was certainly inadequate; it took over six hours to get the men out of the tank. The time of the accident could be significant. By 5 p.m. the painters may have been overly tired. Perhaps someone shut off the blower, forgetting that workers were still in the tank.

In the NIOSH report no. 80-106 to which I have already referred, there is an account of a fatality that occurred as a result of cadmium poisoning. Two men were using a torch to cut 26 cadmium plated bolts in an underground vault $18 \times 6 \times 5$ feet with a 24 inch diameter opening. Neither man wore a respirator, and no ventilation was provided until after the torch work, which created a heavy blue smoke, was completed. At the end of the shift, one man became nauseated and the next day he complained to his doctor of fever, chest pains, coughing, and a sore throat. Three days later he was hospitalized and he died of coronary infarction and corpulmonale less than two weeks after exposure to the smoke. The second worker developed pneumonia but eventually recovered. These delayed reactions to cadmium poisoning are considered typical by medical authorities.

Another accident recorded in the same report occurred when a man was using a torch to heat the metal around the manhole of a tank so that two workmen inside the tank could straighten out a bulge in the metal. The job required that the manhole cover be in place and so they used welder's oxygen to avoid suffocation. It was blown into the tank through another opening until someone noticed that the hair and clothing of the workers in the tank was burning vigorously. One worker managed to climb out, but the other lost consciousness, and it was necessary to invert the tank in order to drag him out. He died shortly thereafter; the other worker recovered after several months in the hospital, and one of the rescuers suffered burns on his hands.

Carbon dioxide and carbon monoxide are produced during gas welding and at the same time the percentage of oxygen in the air may be severely depleted by combustion, displacement, or both. Enclosed spaces should be adequately ventilated while welding or cutting are in progress. In addition, respirators will be needed if the space contains toxic gas or vapor.

Even open pits may contain a dangerous accumulation of gas. Decomposition of organic matter releases methane, fermentation produces carbon dioxide and, if the site has been filled with industrial refuse, many different vapors that were trapped in the soil may collect in a pit. Gases that are heavier than air can collect in pits and remain there even though there is no visible evidence of their presence or apparent reason for them to stay in the pit and not simply dissipate into the atmosphere.

Two contractors that have strong safety programs were negligent in

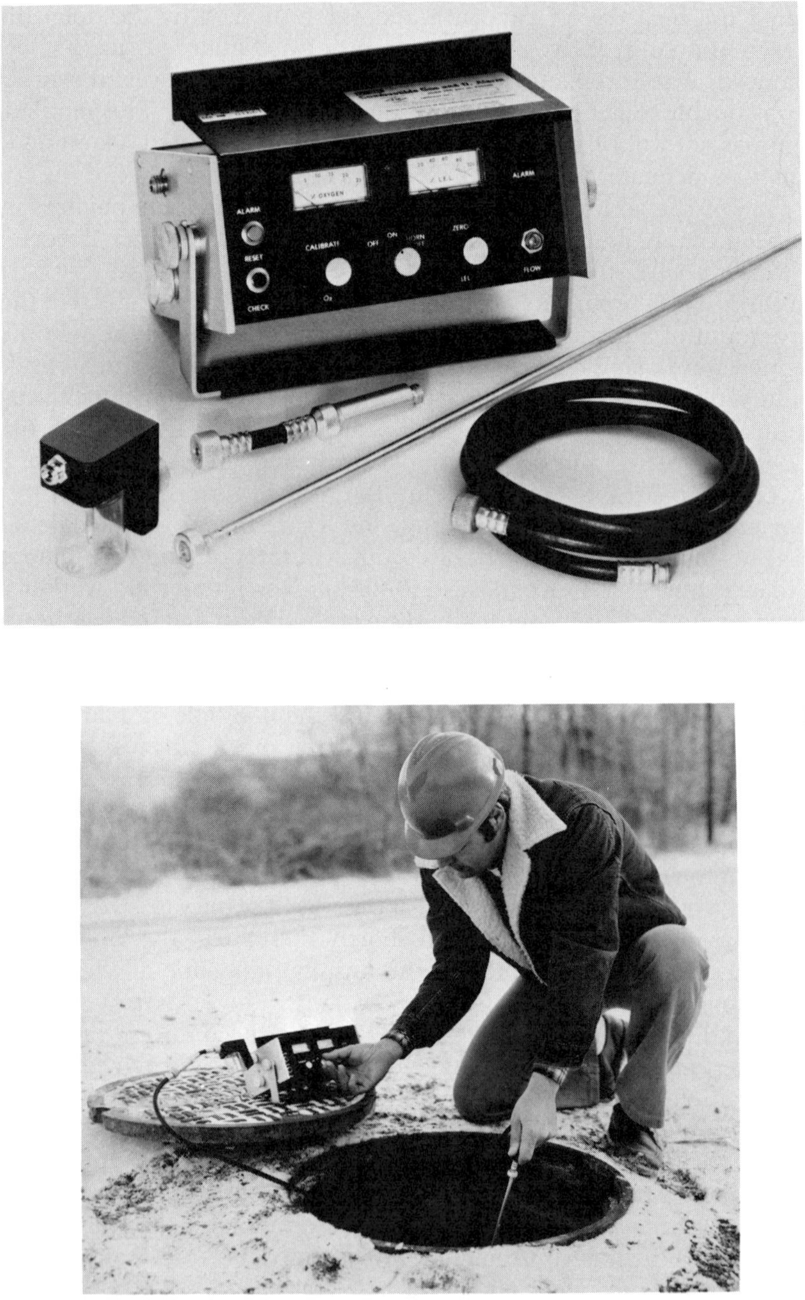

Figure 7.3 Portable equipment is available to measure the percentage of oxygen in the atmosphere in an enclosed space and to determine whether the concentration of a gas in air is approaching the lower explosive limit. MSA's Model 260 assembly is shown in use and the individual parts are illustrated by the adjacent photo. (Pictures courtesy of MSA, 600 Penn Center Boulevard, Pittsburgh, Pennsylvania 15235.)

allowing workers to enter a well excavation at a large power plant development. They did not anticipate that methane would be present and, when it exploded, two workers were severely burned. The workers collected compensation from their employer, the well contractor, and then sued the general contractor for negligence because he had participated in planning the safety program. The courts refused this claim, but meanwhile the case was considered by two courts. The whole affair could have been avoided by a simple check of the air quality for the presence of methane.

Figure 7.3 shows a portable combustible gas and oxygen alarm, model 260, made by Mine Safety Appliances Company, 600 Penn Center Boulevard, Pittsburgh, Pennsylvania 15235. In operation, a sampling tube up to 50 feet long is passed into the enclosed space, and the instrument draws in a sample of the air. If it contains less than 19.5% oxygen or explosive gases in excess of any selected percentage of the lower explosive limit, an alarm sounds. In addition there are meters that permit the actual oxygen and explosive gas content of the air to be measured independently.

Another very useful, reliable, and portable instrument is shown in Figure 7.4. It is Mine Safety Appliances Company's Samplair Pump, Model A. The slender tube that projects from the end of the instrument is one of about 140 different indicators, each made to detect a particular air contaminant and to indicate the concentration in the air. The tubes are sealed glass capsules, containing a chemical that changes color in the presence of the specific material for which a test is to be made. The user snaps off the seals at the ends of an appropriate tube, inserts it in the hand-held pump, and draws a measured amount of air through the tube. The test result is then immediately determined by reference to a calibration card or a scale on the tube itself. When the atmosphere is likely to be too dangerous to permit the test instrument to be hand-held in the space, a 50 foot long tube may be inserted between the glass capsule and the hand pump. This instrument is very easy to use and tests take only a minute or two. It is not, however, suitable for checking some explosive atmospheres. Both the instruments shown are therefore needed on typical jobs where work in enclosed spaces and exposure to toxic gases are involved.

Whenever a worker enters an enclosed space he may be quite literally putting his life in the hands of the people outside. If anything goes wrong, he may be unable to help himself and must have trust in others to provide all possible safety precautions. He has a right to expect that they will be able to get him out before he dies or is permanently incapacitated if any unexpected trouble occurs. Fast action is of utmost importance. Therefore, the whole operation should be carefully planned before it is begun.

I recall hiring a subcontractor to clean an underground fuel oil tank about 8 feet in diameter and 30 feet long. There was one manhole, and a heating coil had to be removed through it to complete the work. The contractor had been in business for some years, his insurance papers were all in order, and so I assumed the job would be a routine operation. An

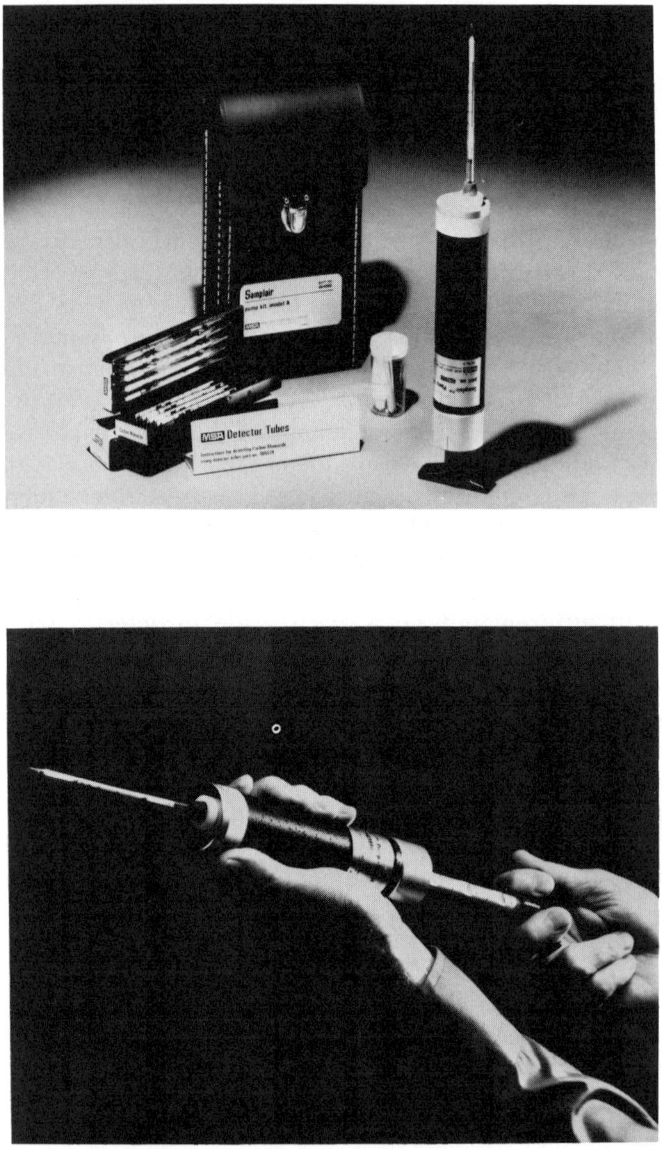

Figure 7.4 Air contamination test equipment. If operations may release toxic gas, vapor, or mist into the air, it is important to know whether the concentration of the contaminant approaches or exceeds the threshold limit range established by OSHA or other regulatory agency. One of the many useful test instruments is MSA's Samplair Pump, Model A, designed to draw air through a wide variety of disposable sampling tubes. Color changes occur if the air contains the contaminant to which the tube is sensitive, and the concentration present can be read against a calibration scale. (Photos courtesy MSA, 600 Penn Center Blvd., Pittsburgh, Pennsylvania 15235.)

hour after the men started work, our foreman called and asked that I come to take a look at what was going on. Two men were in the tank, one man was on the floor shouting down to them through the manhole, and the heating coil was stuck in the manhole so they could not get it in or out. There was no room for the fresh air duct to get past the coil. The men in the tank were almost invisible against the black background; they were covered from head to foot with sludge from number 6 fuel oil. Their foreman had disappeared. Fortunately, we could get a fork lift from another location a few blocks away, and pulled out the coil. The men left and we never heard from them or their boss again.

I strongly recommend that an enclosed space permit system be a part of every safety program under the control of a designated gas free engineer (GFE). Before any work is done in an enclosed or confined space, the responsible supervisor, usually the foreman of the workers who will do the job, is required to obtain a permit from the GFE, stating what is to be done, where, when, and what precautions will be taken to protect the workers. The GFE should check the work plan and discuss any doubtful points with the foreman. He should assure himself that the workers have been trained and that the necessary safety equipment will be at the job. When the GFE and the foreman are in agreement on all points, both should sign the permit and have it posted at the job location just before work begins. When the job has been completed, both should sign it again as a record that the work was safely completed. If trouble developed, it becomes part of the accident report.

Different companies use different forms, depending upon the type of jobs they usually undertake; the form need not be long. It is the quality of the planning that it initiates that is important. The mere fact that two people formally accept responsibility for having thought of the safety hazards and have signed the permit helps avoid oversights. See the sample permit in Appendix J, which was based on NIOSH Bulletin 80-106.

Permits must carry time limits. Often they are valid for only one shift. They should never be allowed to become a mere formality, good for several jobs or for an indefinite time. Their value is in the specific details they include and the recognition that every job will be different from every other seemingly identical one. Jobs are usually classified to indicate the probable degree of hazard.

Class A jobs are immediately dangerous to life or health because there will be oxygen deficiency, an explosive atmosphere, or toxic substances will be present.

Class B jobs are less dangerous than those in the A category, but they still present significant hazards to life or health if adequate precautions are not taken.

Class C jobs are those that involve work in enclosed or confined spaces, but after careful consideration it has been agreed that only normal working procedures will be needed. Work that may seem at first glance to be obviously class C may turn out to be a class A or B project after further consideration.

Most permits include a list of valves, switches, or equipment that is to be tagged and locked while work is in progress. Simple tagging is not recommended; both tagging and locking are advisable.

When normal working conditions are expected to prevail, but there is the possibility that workers may be exposed to varying concentrations of dangerous substances they may be supplied with disposable badges that measure their cumulative exposure to specific chemicals. When the total exposure reaches a predetermined amount, a portion of the badge changes color. If the potential contaminant is odorless, a badge of this kind will provide a visible warning that work should be interrupted until the condition is corrected.

Permits should be specific with regard to all the safety gear that is to be provided and it should be checked off before work begins. The foreman should initial the list and keep it until the job has been completed. All tests of air quality should be recorded and kept with the permit. Be sure that work lights for jobs where flammable or explosive vapors may be present are approved for that service; 29CFR 1926.40(i)(4) requires that portable electric lighting used in moist or hazardous locations such as tanks shall operate at a maximum of 12 volts. Special flashlights are made for use in explosive atmospheres. For emergency use there are also lights that require no batteries and generate no heat. Chemical action inside a sealed tube is initiated by flexing and shaking the device; the light is the result of chemical action. Once started, the tube will continue to glow for several hours.

When the job is about to begin, record the names of the workers and, if they required special training, the name of the person responsible for instructing them should be part of the record. Sometimes the instructor is asked to sign the permit. Include the names of standby workers who may be required to give first aid or to assist in getting workers out of a confined space in an emergency. Be sure all people involved in the operation are familiar with their assignments and ready to proceed.

All this may seem to involve a great deal of paper work, but anyone who has prepared accident reports will realize that the work permit preparation is child's play compared to the investigation of an accident. Those reports, and the litigation that may follow can take hundreds of man-hours. Prevention is easy by comparison.

While work is in progress in a confined space, be sure that someone is monitoring the air supply to the workers. Air must be blown into the space fast enough to guarantee there will be no oxygen deficiency anywhere, or a portable air supply carried by the worker is required. The GFE must consider what type of respirator is needed, if any, and he will often want to confirm his selection by discussions with the manufacturer to be sure there have been no recent developments of which he should be aware.

For example, the Department of Health and Human Services, National Institute for Occupational Safety and Health (NIOSH) issued an emergency information bulletin in January 1982 on the danger of using self-contained breathing apparatus (SCBA) in hazardous environments in subfreezing weather. The principal points covered were:

Small amounts of moisture in the air supply may freeze and cause failure of the apparatus.

A nosecup is needed to avoid faceplate fogging, and other accessories may also be recommended by the manufacturer for cold weather operation.

SCBA's can freeze quickly if their use is interrupted. Keep the equipment warm until it is needed again.

Alarms can fail; check the remaining service time visually.

Exhalation valves can freeze in the open or in the closed positions. Manufacturers can supply specific instructions for coping with this problem.

After washing faceplates and breathing tubes, dry them thoroughly. Then keep the equipment in a warm and dry location.

7.8 Testing Mechanical and Electrical Systems

Some accidents that occur while mechanical and electrical systems are being tested are caused by defects and others are due to improper testing methods. Inadequate testing can result in injuries to occupants of a structure after it has been accepted by the owner.

Some degree of danger is inherent in testing work, simply because tests are made to find defects. Anyone who opens a valve on a water line with complete confidence that water will come out probably hasn't done much testing. It is safer to be prepared for steam, compressed air, gas, or fuel oil to flow than to anticipate only the proper fluid will come out of a potable water tap.

To speed up the job, temporary quick fill lines are often installed to interconnect potable and closed recirculating water systems. When these lines are simply valved off and not removed, they constitute a safety hazard for as long as the building is occupied. Toxic chemicals may be added to a closed system years after the building is completed, and backflow through a "temporary" line is always a possibility. Before a system is considered to have been properly tested, all such connections should have been removed and not simply valved off.

Anyone who has seen the muck that collects in a typical cooling tower sump, or has fished dead birds out of one, will appreciate the health hazard that exists if condenser cooling water can be siphoned into the potable water system. Proper testing of such a system should include inspection of the float valve discharge; a submerged inlet is a common and dangerous cross connection. The end of the inlet pipe should always be safely above the overlow connection. Many other common cross-connections may be found in typical buildings, in violation of sanitary codes and in spite of the inspection work that is presumably done by local authorities.

The U.S. Environmental Protection Agency's Water Supply Division has published a 57-page "Cross Connection Control Manual," EPA#335. It covers the basic principles. A somewhat more technical manual is available from the American Water Works Associations, 6666 West Quincy Avenue, Denver, Colorado 80235 under the title "Cross Connections and Back Flow Prevention." Both publications are worth reading and include case histories of illnesses caused by cross connections that allowed bacteria, water treatment chemicals, gasoline, and other contaminants to enter potable water lines.

During testing and trial runs, most electrical system defects will probably be found but, if there are some poorly made connections, it may be some time before the resultant overheating will cause the insulation to fail completely. I am not likely to forget touching a poorly supported connection box that contained a wire on which the insulation had deteriorated. When I touched the box, a short circuit occurred and burned a hole completely through the metal before the circuit breaker opened. It took some time for the burned hand to heal.

Power transformers should be capable of withstanding surges of current that will open circuit breakers but, if a dead short exists when a circuit is energized, electrical forces between the windings may tear them apart, causing internal arcing and if the equipment contains flammable transformer oil, a fire may result. I have inspected enough of the wreckage of transformers to have a healthy respect for the damage a failure can cause. Remember that a unit may have been perfect when shipped, but it is possible for damage to occur between that time and the first in-service use. Check current transformer (instrument transformers) carefully before energizing a circuit in which they are installed. Dangerously high voltages may be expected across the secondary terminals if ammeter connections are incomplete.

In any building, a third or more of the three phase motors will probably run backward when first energized. Be prepared for any hazards that may result, such as loosening of pump impellers, or unexpected motion of belt drives.

It is important that all 120 volt outlets be tested to be sure they have been properly wired. Reversing the ground and neutral wires creates a safety hazard, even though a test light will burn when plugged in.

To avoid accidents, circuits should not be energized until the electrical contractor has completed his tests and made all necessary modifications.

Workers should use a lockout system as soon as electric power, water, gas, or other fluid is available at a piece of equipment on which they are working. Every worker who might be injured if a switch or valve were to be actuated without his permission should attach his own tag and padlock to it. Safety supply houses sell locking devices to which as many as six locks may be attached. No worker should rely upon another's lock for protection; his co-worker may be called off the job and take his lock with him without notifying anyone.

Pressure testing should be done with water whenever permissible. If freezing weather is anticipated and water left in a system might cause damage, it is safer to add antifreeze than to use compressed air or nitrogen because if a connection or fitting breaks, the expansion of compressed gas can cause broken lines to whip about, and the blast of air or gas can drive loose dirt from the piping or pick up material from the floor. By comparison, when a system is being tested with water pressure, a break causes the pressure to drop very quickly. Of course there may be water damage, but unless this is an overriding consideration, compressed gas testing should be avoided. Never use pure oxygen or acetylene for testing when air or nitrogen should be used; leaks can cause fire and contact between oxygen and oily material may result in an explosion. Nitrogen is generally considered an acceptable gas for test work, but if it escapes into an enclosed space it may displace enough of the atmospheric oxygen to endanger workers. See Section 7.7, which deals with working in enclosed spaces.

Specification writers have been known to require hammer testing of pipe welds while the system is under pressure. I recommend challenging such specifications because they can endanger workers and damage equipment or piping. Hammering on a piping system that is full of water can send shock waves through the lines, greatly increasing the momentary system pressure. It is an extremely punishing test and some manufacturers will consider it grounds for voiding the warranty on piping or equipment. Hammering on a system that is full of compressed gas will endanger the worker and anyone else who may be in the vicinity if a break occurs. If the designer insists that a hammer test be used, I recommend that it be done only when the system is empty and that the designer be required to state the weight of the hammer that is to be used and the way it is to be used.

If a leak is found, relieve the pressure in the system before attempting to make a repair. Welding on a pressurized system can cause hot metal to be spattered onto the worker or onto flammable materials, particularly if there is air pressure in the system.

During pressure tests, if it is necessary to remove the regular pressure relief valves, they should be replaced during the test with temporary ones set just above the test pressure. Always be prepared for the controls on the pump, gas container, or compressor that is used to pressurize a system to malfunction. In many instances a dangerously high pressure may be applied before anyone realizes it. This is apt to happen when water is used and the system is completely full. Water is almost incompressible and forcing even a small amount into a system that is already full will rapidly increase the pressure. Also, if a system is filled with cold water and pressurized, if the water temperature is raised by exposure to summer weather or the sun, the pressure may build up enough to split the pipe or damage equipment. It is under such conditions that the temporary relief valves are most important.

Before any system is tested, consider the possible types of failures that may occur and keep people out of dangerous places. Remember that overspeed controls may not work and consider the no-load speed that may be reached by all motors. A few of them may be series wound and will reach dangerous speeds if a shaft coupling was not properly attached. I recall handling the parts of a cast iron pulley that disintegrated when a series motor was belt-connected to a load and the belt broke. One piece of the pulley flew through the air, clipped off a wire that supported a light, went through a window, and landed on the seat of a car about 200 feet away. Luckily nobody was injured. At high speeds, commutator bars may be thrown out of a motor, and it may tear itself apart when the windings fly out of the slots. Such accidents are rare, but when they do occur the potential for injuries is great.

Some systems are subject to vibration when a defect is present. For example, a loose vane in an air conditioning duct may flutter and shake loose the rest of the adjacent sheet metal work. Anything supported by anchor bolts that are not properly set may fall if the vibration is severe and continues. Be prepared for belts to snap, and for hoses to split or become detached from their connectors. Remember that a "stop" button may not necessarily stop a motor; have someone stand by the circuit breaker or disconnect switch until the stop button has been proven reliable.

Small chemical feed pumps are frequently equipped with fractional horsepower motors. They do not look dangerous, and it is easy to overlook them when testing other system components. But, if the pump discharge is clogged, valved off, or includes a check valve that was inadvertently installed backwards, even a small pump will generally be capable of building up a discharge pressure well in excess of 1000 psi. Workers have been sprayed with strong chemicals when such high pressures split a fitting. Chemical feed pumps should always be equipped with either a pressure relief valve or an internal pressure limiting device set to prevent dangerous conditions from developing. In service, chemical deposits often form in or at the end of the pump discharge line. Consequently, if this safety detail is overlooked, an accident may occur several months after the job is completed.

Strong chemicals are needed when dirt must be removed from a piping system. The cleaning solution must be capable of emulsifying oily residue from threading work, and of dissolving temporary protective coatings that may have been applied to prevent superficial corrosion prior to installation. Sometimes fuel oil gets into equipment by accident. Chemicals that are strong enough to be effective should only be handled by workers who have been trained and equipped to avoid injury to themselves or the system that is to be cleaned. The waste solution must be neutralized before being discharged into streams or sewers in order to avoid violation of EPA regulations, and the neutralizing work also must be done properly to avoid precipitating insoluble material from the cleaning solution. No traces of chemical should be unintentionally left in a system; some cleaning solutions that are safe to use when contact with metal is limited to an hour or even overnight can cause serious damage if allowed to remain for too long. Sterilizing of potable water lines is generally done with some form of chlorine. This gas is toxic and extremely irritating to the respiratory system and even when a solid chemical that releases chlorine in solution or liquid sodium hypochlorite is used, the manufacturer's recommended safety precautions must be very carefully followed. After the system has been filled completely with the sterilizing solution, tests must be made to prove that the work has been effective and then all traces of the strong solution must be removed. At every stage of the work, people must be protected. They may attempt to drink the water and even though the taste will be repulsive, irritation can occur because it is unlikely that any source of other water will be readily available in the building that could be used to rinse out a person's mouth. Eye protection is also imperative.

I recommend that sterilizing work be done by or under the continuous supervision of a certified laboratory that is acceptable to the local department of health. When the work is completed, a laboratory report should always be obtained, certifying that the tests were made and that the job was done in a professional manner. That is not to say that an employee of the contractor could not do the job just as well as a laboratory technician; the testing is not difficult to do. But, if an individual becomes ill after drinking water from the system, the laboratory report would be acceptable proof that the illness was unrelated to the water, whereas a statement by a contractor would be challenged and would have little status at a trial.

Be cautious about following specifications that list specific chemicals and concentrations; many of them contain typographical errors or have been written by people who copied them from a text and did not fully understand the importance of accurate transcription. A common source of misunderstanding is the use of the abbreviation "M" to indicate a quantity of water. For example, a specification may call for using a certain amount of chemical per M gallons of water, and in the water treatment industry the abbreviation is sometimes used to indicate 1000 gallons, based on the Roman numeral M being the symbol for 1000. It is also used to designate 1,000,000 gallons. Therefore, if the specification is not clear and the worker who prepares the solution does not know enough about his work to understand what is meant, it is possible to make a serious error. Similarly, the difference between a chlorite, a chlorate, and a chloride will not be clear to anyone who does not customarily handle chemicals and he may order the wrong material from a supply house.

When water is poured into a concentrated solution of a strong caustic or a strong acid, heat is released and spattering is likely to occur. The same effects are produced when water is poured onto powdered or granular caustic and when a strong acid is mixed with a strongly alkaline material. The heat may cause severe burns, and spattered chemical may do irreparable damage to the skin. Strong chemicals should always be added to enough water to keep the temperature under control, and the additions should be made slowly. Workers who have not had experience in making up solutions often get into trouble because they forget or are ignorant of the heating effect. I recommend that whenever specifications require that chemical cleaning or chemical tests be made that a chemist or lab technician be present to check and supervise the work.

Fire alarm and security systems should be tested when only completely trustworthy people are present. They are necessarily equipped with shut-off valves and switches that would be of interest to potential criminals, and the fewer witnesses there are to tests the less opportunity there will be for the wrong people to know how to shut them off.

7.9 Painting

Painters are most apt to be injured by falls, possibly because they are often tempted to overreach rather than to move ladders. There is little that I care to add to the comments I have already included in Section 4.2.

The dangers associated with painting in enclosed spaces have also been mentioned in Section 7.7. Spontaneous combustion due to slow oxidation of oily rags was mentioned in Section 6.4.1. But there are some hazards that are more specific to the painters than to other craftsmen. Lead poisoning should be avoided when old surfaces are being sanded. Respirators should be worn and dust that might contain lead should be washed from the hands before eating or smoking. New coating systems may involve the use of volatile solvents with which workers are not already familiar. The old saying that "paint is paint" has not been accurate for many years, and is becoming less true every year. Some solvents can pass into the bloodstream by diffusion through the skin even when there is no visible injury. Even some protective gear such as rubber gloves may not be effective against vapors that are irritants or against some solvents. When gloves are needed it is important to be sure they are impervious to the material to be handled. No one type is best for all exposures.

High pressure airless spray guns are often capable of injecting paint through the skin and causing serious injury. In 1981 a navy yard worker was spraying a paint that contained lead when a high pressure hose leading to his equipment broke and injected paint into his hand. Surgery was necessary to remove as much of the paint as possible, but an estimated 10% of the material remained in his system. Within about two weeks he had to be treated for lead poisoning, the symptoms of which include dizziness, nausea, pain, and weakness.

Keep paint off safety belts and lanyards. If it penetrates either cotton or synthetic fiber gear and hardens, it can cause fibers to break by restricting their ability to flex freely. Paint pigments and solvents may also damage fibers by chemical action.

Epoxies are a class of products that have many useful properties that can be modified by the addition of different curing agents and solvents. They are so important in today's industrial environment that the U.S. Department of Health and Human Services has released a booklet about them, HEW Publication No. (NIOSH) 76-152 under the title "Epoxy Wise is Health Wise." Among the special health hazards that may be associated with this class of compounds is sensitization. After working for years with various chemical curing agents, one chemist suddenly swelled up like a balloon, his skin erupted in blisters and was covered with pustules. He recovered from this severe illness, but if he even walks into a laboratory where there are open containers of any of the chemicals he used to handle so freely, he gets an immediate and violent reaction. He has become permanently sensitized to a whole group of chemicals.

Some people are affected by only certain chemicals and when they do not find a mixture irritating they become careless about washing up and keeping it from contact with their skin. Then they may suddenly be exposed to a slightly different mixture, with immediate results that may include permanent injury. Chemicals that have apparently identical functions in a formula and are considered interchangeable by manufacturers may have very different effects on users. The only safe rule is to keep epoxy mixtures from contact with the skin, and it is obvious that care should be taken to keep food free of contamination. In observing this precaution, many people fail to consider residual material that is on gloves. They take off one glove, and then use the bare hand to remove the other glove, thereby contaminating one hand. Finally, they may stuff the gloves in a pocket, getting the chemical onto their clothes and anything that may be in the same pocket. The correct procedure is to wash up thoroughly before removing the gloves, so they are both clean. Then they may be safely removed and set aside to dry.

Rings and wristwatch bands can trap irritating materials against the skin. It is better to avoid wearing them while working with any product that might be injurious if allowed to remain in contact with the skin for a considerable length of time.

When considering the hazards to health created by the use of any product, its flammability, the effects of vapors on the respiratory system and skin, and its oral toxicity are all important. The hazards may not be confined to the painters; other workers may be exposed while coatings are being applied or while they are drying, particularly when ventilation is poor or when forced ventilation is stopped when the painters finish their work.

Evaluation of the hazards associated with the use of any commercial product may be a highly complex project requiring the expertise of chemists, industrial hygienists, fire prevention specialists, and experts in the choice of protective gear. Many proprietary products are mixtures of ingredients in proportions that are not disclosed, and some components may be identified on the label in such general terms that even the experts cannot be sure whether the product is dangerous to use or just moderately hazardous. Some manufacturers are very cautious in labeling; they lean over backwards to be sure they cannot be accused of failing to warn the user. If taken literally, some warnings would make one hesitant to use the product under any conditions that could not be completely controlled. On the other hand, some manufacturers do not want to jeopardize sales by including any more warnings than are absolutely mandatory. Vague language is often used. "Use only in a well-ventilated space" appears on many labels, but what does it actually mean? Is forced ventilation required, and if so, how many cubic feet of air should be handled per gallon of paint used? The same warning may merely indicate that the windows should be opened.

I recommend that safety engineers obtain application instructions from manufacturers, and that they be read carefully to see whether they contain any indication that health hazards may be involved. If so, a call to the vendor's plant may be the best way to get full and realistic information. Ask to speak to the chief chemist or the head of the industrial hygiene department; someone at the plant where the product is made will know whether it has to be handled with unusual care and, if the field conditions are described to him, he can be expected to make some practical suggestions.

When a work space is cleaned, remember that the dust created by sanding, use or rollers, or overspray may be explosive if mixed with air. Never dump dusty material into a burning barrel, regardless of whether there is a fire in it or not. If the paint contains a flammable solvent, do not permit anyone to use an electric drill or power sander where an explosive atmosphere may be found. Most power tools create sparks and are as dangerous as smoking when flammable vapors can pocket in a poorly ventilated pit or compartment.

Unusual painting problems may be involved when a structure is exposed to industrial fumes, wet and dry marine (tidal) action, or unusually great temperature fluctuations. Painting a smokestack is different, for example, from painting a house and it is important to recognize that one may be tempted to underestimate hazards even when he realizes they exist. Special skills and equipment may be required to prepare surfaces for coating, and industrial coating systems often require more skill to apply than the inexperienced contractor anticipates. Nobody tests the film thickness on a house but, if a chemical resistant coating of a given minimum thickness is specified, the painting contractor should anticipate that the thickness will be checked before the job is accepted. Consequently, when a contractor underestimates the cost of doing a job and sees that he is going to lose money on it, there is a great temptation to cut corners and allow workers to take risks of injury. Safety nets, belts, lanyards, life lines, respirators, and similar safety equipment costs cut into profits if the need for them was not anticipated. Bidding for jobs they are not prepared to complete safely is therefore one reason contractors have accidents. Owners are shortsighted when they accept a bid that is abnormally low. They must anticipate added inspection costs, delays, poor results, and accidents that may involve them in litigation even though they may ultimately be absolved of blame.

7.10 Inspections

This book is primarily about worker safety in the construction industry, but I cannot ignore the public interest in construction safety and the influence of public attitudes on the industry. If a crane falls, pedestrians as well as workers may be injured. Facade defects are a concern of builders and public alike. Certainly fire hazards created by poor codes or failure to enforce realistic ones cannot be shrugged off by owners or contractors, but it often seems that the public interest conflicts with the profitability of projects. In order to be competitive, builders and owners concentrate on meeting the letter of the code, regardless of whether the end result meets the best needs of the public.

Isn't it obvious that the only interest of typical owners and contractors in inspections is to get an approval of the work that has been done? The responsibility of an inspector to find and require correction of defects that could cause fire or structural damage is unlikely to be appreciated while a project is under construction, but he is certain to be blamed for overlooking anything if there is loss of life after the structure is completed. This is understandable; nobody wants to do work over, or to be told of a code violation. Delays cost money, and many owners are more interested in speed in getting the certificate of occupancy than in the things the builder does to expedite the process.

I am never surprised to learn that inspection work was poorly done, or that approvals were speeded up by questionable or clearly illegal means. Rather, it is surprising how many inspectors are diligent and thoroughly honest. Theirs is a thankless job. Even when a fire or other loss occurs, it is their oversights that are publicized; the work they did well is taken for granted. Still, if the public is to get safer structures, the quality and consistency of the plan review and inspection work must be improved. The problem of achieving this improvement is a difficult one.

The enlightened self-interest of owners could be instrumental in this regard. Consider the disasters that occur every year as a result of poor cutting and patching work that allows fire and smoke to spread unchecked. Recall jobs that collapsed because they were improperly designed or built and the defects were not observed. They cost owners dearly, but news of a relatively small percentage of jobs that are disasters does not seem to deter others who want to cut corners.

No fast turnaround is likely, but in the long run I think it will be public reaction to unsafe construction that will make the industry more consistently aware of public safety. There are some signs that this trend is beginning.

Some large corporations require that a safety engineer visit the premises before they schedule a meeting in a hotel or restaurant. I have been asked to check a few locations to see whether there were sprinklers, adequate exits, and clean kitchens. If inspections were a common preliminary to signing contracts, hotel managers and owners would soon pay more attention to safety precautions. Alterations made to meet the demands of large organizations would benefit, of course, the general public as well. Individuals are conscious of the value of smoke alarms in their homes; business executives could help if they would check the safety of facilities when they arrange for non-business affairs such as club meetings. Just a few simple requests for rooms below the eighth floor can remind hotel managers that people are becoming interested in fire safety. Inquiries of this nature by business executives are particularly apt to get attention. People who travel on business are on the road more than typical vacationers, they spend expense account money more freely than many other people, and they return to the same cities often enough to be able to compare and choose between accommodations. Bad publicity about one unit in a hotel chain could affect business in many other cities. In anticipation of increased public awareness of safety, owners may begin to pay more attention to the quality of buildings while they are being erected.

Some owners already supplement the work of city inspectors by retaining consultants to observe work in progress with sufficient care to be aware of oversights by city inspectors or plan examiners. They consider the consultant's fees good insurance that they will get the quality of construction for which they are paying. An independent consultant can also feel free to advise an owner when it would be worth the extra cost to do a better job than the local code demands.

Building contractors lose millions of dollars worth of business every year because owners delay correcting code violations. Court calendars are crowded and, as long as money is not spent for necessary alterations, it can be earning profits elsewhere. Tenant associations and businesses that occupy leased space could step up pressure on reluctant landlords to install smoke alarms, fire doors, panic bars, and whatever other deficiencies have been found by inspectors. Tenants are as much entitled to a safe building as to one with adequate heat, light, and water. More would be done if more were demanded. If speculators anticipated trouble renting unsafe buildings, there would be fewer of them built.

The good name of a building management firm is one of its most valuable assets. People prefer to rent from agents that are reputed to be good to their tenants. Every night the TV news features some organization staging a demonstration. Conservationists and consumer groups have become more active every year. Public indignation and publicity could be effective in reducing hazards in buildings, to the benefit of the public and to the long-range benefit of the building industry.

If the public spotlight is focused on the weaknesses of the construction industry, the people involved might begin to act more responsibly. People in the business might even find that by and large they can trust one another and that much of the bonding and boiler plate in specifications could be eliminated. Investors commonly call brokers and place verbal orders to buy or sell securities that are worth as much as a small building. Imagine the chaos that would result if every investment transaction had to go through the same title search, negotiation, and closing procedure we have to follow before beginning a construction job. True, there are swindles in the financial community, but I doubt they are more prevalent than in the construction industry in spite of all our checks and doublechecks. We are hamstrung with inspections because the industry has not learned to manage its affairs on the basis of mutual confidence and integrity. Therefore, to the extent that inspections can be eliminated or made an unnecessary formality because the work will have been done well without having to be policed at every stage of the job, it will contribute greatly to the prosperity of all concerned.

8

Finishing the Job

8.1 "Changing of the Guard"

There are times when pomp and ceremony are appropriate. Every morning at Buckingham Palace, the traditional changing of the guard takes place; at a certain time responsibility for whatever the guard is expected to do that day is formally transferred, and there is no question about who is in charge at any time. In the Navy, the traditional "I am ready to relieve you, sir" and the few set phrases that follow have been accepted for generations because they serve a useful purpose. The officer in charge at any moment knows it, and he knows exactly when he has been relieved of his responsibility.

When a construction job nears completion, the "completion" date is hard to establish; there will inevitably be some items left on the punch list that will not be completed until the building is occupied. Some touch-up painting, a little balancing of the heating or airconditioning system, perhaps some noise in the pipes or duct work, finishing of the landscaping, or any one of a hundred minor items may be awaiting completion. But from the viewpoint of the safety engineer, the formality of a transfer date has value. As of any given moment, it should be clear who is responsible for security and the prevention of accidents. In this chapter I mention some typical items to be considered in the transition period; not all of them need be transferred at the same moment, but for each item there should be a time at which responsibility is transferred. Accidents can happen when two people both think the other is taking charge of an operation and neither is doing it.

8.1.1 Training in the Safe Operation of Mechanical Systems

In Scotland there is an old saying "He came in with the bricks," meaning, of course, that someone had been at the job since its inception, and that he is completely familiar with every detail. Some owners still choose their chief operating engineer as soon as construction begins, and have him live with the job until it becomes his responsibility to run it. Other owners seem to think any engineer can walk onto a job and know all about running it immediately. I think it is poor judgment on the part of any contractor to turn over a mechanical system to an engineer without being reasonably sure he knows how to operate it, and it is equally unwise for an engineer to accept responsibility without getting all the information he needs about it.

I recall one classic example of failure of the owner, the contractor, and the operating engineers to recognize the need for an indoctrination period. It was more a case of good luck than good management that the engineers and the system survived. It was a total energy system in a city where reliable power from a utility was available, but for the sake of anticipated energy cost savings, the plant was isolated without even a standby connection. Four diesel generators provided electric power, and each engine was connected to a steam generator through which its exhaust gases were passed. The steam was used for heating in winter and to operaate absorption airconditioning equipment in summer. An elaborate automatic control system was intended to eliminate the need for watch engineers; one resident chief engineer was expected to run the job by himself. Naturally, every engineer walked off the job as soon as he found out what was expected of him and every replacement had to start from scratch. I have been told that the promoter declared himself bankrupt and the project was abandoned for a time before being reorganized.

Many operating engineers can recall comparable experiences when they had no opportunity to familiarize themselves with a system before becoming responsible for its safe operation. But those are memorable incidents because they were stressful; there were unknowns that could have caused accidents. At a new job, the engineer should insist that he be given time to become familiar with the equipment in his plant before being responsible for its operation. The contractor and the engineer should agree on the time responsibility is transferred, and thereafter the contractor should carefully avoid making any changes or adjustments without the engineer's knowledge and consent.

Safety engineers should be sure that instruction manuals cover the precautions to be followed in operating all equipment. When manuals have been reviewed with the operating engineers and delivered to a responsible party, it is appropriate that he get a receipt and an acknowledgment that the material has been adequately explained. A certain amount of formality and protocol emphasizes the importance of the manuals; they are more than just catalogs.

If safety gear is supplied by a contractor, he is entitled to get a receipt for it signed by the person to whom it was given and also by all those who were instructed in its use. Specifications may require that operating personnel be trained in the use of safety gear, and to do this properly can take an appreciable amount of time. On one job for a public utility that operated many buildings, specifications required that their employees be instructed in the handling of chemicals and making water treatment control tests. To do this properly took about two days, and classes were attended by some supervisors as well as the men who were to be trained. All agreed the course was a good one, but within a month the utility asked that some "new" employees be trained because so many of the original trainees had been transferred to other work. More manuals had to be prepared, and complying with the request was costly. On future jobs, the water treatment company insisted that the number of people to be trained be specified and that additional training be considered professional consulting work. The idea was accepted; the utility realized that training should be considered just as much a part of the job as mechanical components.

Note that I have suggested that workers be required to sign a statement that they have been instructed in safety matters. Some people may object to signing anything; if this may be a problem, consider having some simple diplomas printed, certifying that the training was completed. The signature of the instructor and acceptance of the diploma by the worker whose name appears on it is almost equivalent to a signature by the worker and is good evidence that the training was provided if copies of the certificates are filed by each worker's employer and by the contractor. At the end of a day or two of training, most trainees also like to get some tangible evidence that they attended the course.

8.2 Inspection and Acceptance Reports

In the previous chapter I commented on the importance of tests and inspections. At the time a job is completed, the records of all such work should be complete. If, for any reason, an inspection was voluntarily waived by the owner or anyone on the design team, the contractor should make it clear that failure to inspect is tantamount to acceptance. In Chapter 6, Section 6.6, I mentioned that investigators reported that the architect and design engineer had not considered it necessary for them to make a final inspection of the space truss that later collapsed at the Hartford Coliseum. After the fire at Stouffer's Inn in Westchester County, New York, investigators could not locate certain important inspection reports that were or should have been filed by the local inspector. He had died before the fatal fire occurred. When anything goes seriously wrong, seemingly unimportant details often become vital links in an investigation. If a report is missing, it is sometimes possible to repeat the inspection and, if this cannot be done, it is advisable for the contractor to report the facts rather than to risk the accusation later that he deliberately attempted to cover up something.

A walkaround inspection should be made and reported to confirm that all belt or coupling guards were installed, that tripping hazards were eliminated, guard rails were in place, exit signs were lit, fire extinguisher locations were properly placarded, and that all similar safety features were provided in accordance with specifications. A report of this tour, signed by the owner's representative, protects the installer against charges of negligence or failure to comply with specifications if someone later removes and fails to replace one of the items listed, resulting in injury to someone. I have often been unable to find safety gear that was delivered to a job; small items seem to disappear during the hectic days immediately before and after completion of mechanical systems. Without evidence that they were turned over to some responsible employee of the owner, the contractor may find it impossible to avoid replacing gear that costs an appreciable percentage of the net profit on the job.

If, during an inspection, some items were missing, be sure to get a signature when the omission is corrected.

Give some consideration to the best time to remove temporary safeguards; leave the tape on glass as long as possible. Keep sheds over sidewalks until all overhead work has been completed.

Contractors are usually required to submit working drawings for approval before proceeding with construction, and the general contract often requires the architect to prepare "as built" drawings when the job is done. If he is entitled to receive copies of any such drawings, the contractor should follow up to be sure he is not forgotten. If any trouble occurs and it is necessary to find buried lines, or if any additions to the structure are planned, it is better to have your own set of final drawings than to rely upon the owner to retain them in good order. At the end of the job, while details are reasonably fresh in mind, it pays to take a little time to mark the plans with any notes that could affect the safety of people who may have to do emergency repair work. For example, record the color coding of control wires and piping, particularly if there are any unusual features of the system. Make notes to indicate the meaning of such terms as "primary" and "secondary" systems as they apply to the actual layout. This may seem unnecessary while everybody recalls the details of the job, but a few months later when someone wants to trace lines in a hurry it is handy to find notes that indicate that "on the 44th floor the upper secondary system has a compression tank and the drain valves for condenser water lines are just west of it." If valves have number tags, get a key list indicating what each one controls and where the important ones are located. Imagine that someone accidentally broke a sprinkler head; would the location of the proper shutoff valve be readily found if the chief engineer was on vacation and only the plans were available?

8.3 Insurance and Security

The dates on which insurance policies are terminated and others become effective should be very carefully coordinated to avoid gaps during which important protection is missing. It only takes one day's gap for an uninsured accident to happen.

Responsibility for controlling access to the premises should also be clear at all times. This is particularly important while occupants are moving into a building; outsiders cannot easily be distinguished from tenants or their employees unless some identification and verification system is used. At the end of a busy day, be sure nobody is left in the building who should not be there.

There may be legitimate questions at the end of a job regarding ownership of spare parts and material left over from the construction work. Sometimes all extra tile, carpeting, paint, glass, and similar material belong to the building owner, particularly if the materials were woven or mixed to meet specifications and are not standard off-the-shelf items. To allow a contractor to remove such material places the owner at the contractor's mercy if a small quantity of it is needed to make repairs. On the other hand, an installer needs to have enough material on the job to be sure he can finish it without running back to the shop for more; he should not be penalized for his foresight. These are not safety considerations, but they do fall under the heading of loss prevention and as such are often the responsibility of the safety engineer, along with security problems.

A special effort must be made to control locks and keys during the transition period. Some tumblers may have to be changed if keys cannot all be accounted for. Workers may have left their padlocks on switches or valves, and particular care must be taken to avoid danger to anyone when they are removed by anyone else. The worker himself should remove his lock unless there is a good and sufficient reason for having someone else do it. The lock may have been inadvertently forgotten by the worker, but it is risky to assume that this is the case. Some part of a system may be incomplete, and there could very well be a valid reason for the lock to be in place. The person who left it may not have been able to finish his job and have deliberately left his lock until he could get a missing part or could test the system.

Watch sprinkler valves while people are moving into a building; I recall an incident when the main shutoff valve was found closed several times and although nobody was ever accused of closing it, the suspicion remained that a small fire might have been planned as a diversionary tactic while a robbery was committed.

8.4 Service Agreements

Early in a construction program agreements may have been made whereby a doctor was retained to provide emergency services, an ambulance was advised to be ready if an emergency arose, temporary sanitary facilities may have been rented, or deposits may have been made to guarantee the return of oxygen, acetylene, or nitrogen cylinders. Even pallets may have been charged to the purchaser of the material shipped on them, with the charge being refundable on return of the pallets. Often these charges are far in excess of the value of the item; the supplier considers it his right to make a good profit if the purchaser is careless enough to forget to claim his refund. It may be the responsibility of the safety engineer to cancel only the medical and related arrangements, but the other items can be a needless sacrifice of profit on the job.

When any mechanical system is completed, tested, and accepted by the owner's operating engineer, he has a right to assume that he has been given all the necessary warnings to permit him to operate it safely. There should be no features that a reasonably competent operator would find hazardous. However, he may not realize that there may be important details about which he should be warned. If he is injured because he was not told about them, the contractor and his safety engineer must expect to be blamed.

Sometimes maintenance work should be done by outside service men if it requires special training and equipment. A letter recommending this arrangement should be written to the owner and I suggest that a copy be hand-delivered to the chief operating engineer and explained to him.

I have seen many examples of this in the water treatment field when the contracts require the installer of the heating and airconditioning systems to provide a "water treatment service" for a period of one year after completion of the system. Unless the owner is knowledgeable in such matters and has a clear understanding of what he is to receive, he may find that the "service" consists of an initial dose of treatment chemical in each system, a small supply of additional chemical, and some sample bottles to be mailed to a water treatment company by the operating engineer whenever he wants to have the treatment checked. He may or may not get an instruction book, personal instruction in the application of treatment, test kits, and chemical feeding equipment. Safety equipment such as goggles, respirator, gauntlets, or a face shield will seldom be provided unless the specifications are sufficiently detailed to mention them by name. In the initial weeks of operation, operating engineers often forget about the need for water conditioning until they suddenly find that the cooling tower is full of algae, the condensers are scaled, and the airconditioning system is out of commission. Then, in desperation, he may try to acid clean the condenser without proper connections, training, or equipment. He may add strong biocides to the cooling tower without realizing that many such materials are poisonous and must be handled with great care. The result of lack of foresight in planning for an adequate treatment program or service can be injury to workers and damage to the systems. If the job is a hospital, windowless office building, or computer facility that is very dependent upon reliable airconditioning, the emergency may be a serious one and under such stress operating personnel are apt to forget about normal safety precautions.

In a restaurant where an airconditioning system cut out shortly before the noon rush hour, the operating engineer made an emergency hose connection to get some additional cooling effect by adding city water to the condenser water system, but the result of the cross connection was the contamination of the entire potable water system with condenser water that had been heavily treated with a toxic chromate mixture. Many people could have been made ill if the error had not been corrected by very prompt action; it was a close call.

Many cities now require that nonchromate chemicals be used for condenser water treatment, and most such mixtures are colorless. Consequently, if there is contamination of the potable water by the condenser water treatment, the occupants of the building may not be aware of the health hazard. In prohibiting the use of chromates, authorities have eliminated the valuable warning of contamination provided by the intense color of all chromate and bichromate solutions.

Prior to taking responsibility for operation of any mechanical system that requires water treatment, operating engineers should plan ahead. Often a complete service agreement is the best arrangement, particularly during the first year. It takes the load off the operating staff during a critical period of time and can minimize the risk of injury to them if anything goes wrong.

Other service agreements are also available that may seem more expensive than doing the work with the building's own employees, but safety considerations may be an overriding reason for using them until a smoothly functioning staff can be organized. For example, window washing may be most economically combined with the other janitorial work of the staff, but until permanent employees are trained and ready to take over the job, it is safer to let others do it.

Where extensive lawns and wooded areas surround a structure, training workers to handle and maintain large power mowers and to do tree work takes supervisory time if accidents are to be avoided. Unskilled workers hired off the street can be careless with gasoline, ignorant of the dangers associated with the use of power tools, and unlikely to know anything about tree work. Weigh the value of the time of a skilled supervisor against the extra cost of a service organization that will assume responsibility for the safety of its employees.

Most of the foregoing comments are directed to the building manager rather than to the contractor, but contractors have a continuing responsibility for their work, and if the owner's staff is inefficient, constantly distracted, or there are many personnel changes, it can make many extra visits to the job necessary. They are extra costs and lost profits. To the extent that the safety and loss prevention activities of the contractor can assist the new operating staff in getting the job under control, they can contribute to the contractor's profit as well as the owner's.

8.5 Safety Awards, Ceremonies, and Publicity

When a job has been completed with a good safety record, all concerned have something to crow about. Why should only the bad news make the headlines? When a safety engineer has worked hard to prevent accidents and has been successful, he deserves some recognition. In turn, he may want to award some safety plaques to contractors and subcontractors and the ceremony will probably be pictured in the local paper if the editor is asked to have the story covered.

Safety awards can be combined with a ribbon cutting or grand opening ceremony. Once, when some plant expansion work had been successfully completed, all the construction workers and the plant operators were invited to bring their families to see the changes they had wrought. Youngsters enjoyed seeing their parents operate large machines and marveled at their ability to maneuver a fork truck around tight corners while going in reverse. A few refreshments were served and everybody enjoyed the show. It also helped get across the basic idea that employers did care about safety.

At ceremonies some souvenirs can be distributed; medals with key rings attached, pocket flashlights, or items of more substantial value if the size of the job makes that appropriate. Any tangible item helps remind people of the event, even if it is just some little gadget that is useful enough to keep.

Most contractors expect to stay in business in the same community. Their names are often well known to the local residents, but they are not known as people except to those with whom they have done business. At a ceremonial opening, when everybody is congratulating everybody else, contractors have an exceptionally fine opportunity to be seen as individuals. It never hurts to be known as an employer who cares about the safety of his workers. He is more likely to get job applications from the best workers when there is plenty of work to be done and good men are in short supply. In bad times, if he has attracted and held the best and most productive people, he is apt to fare better in a sharply competitive market. In other words, it can do no harm to be thought of as one of the good guys, and what better occasion than a safety award ceremony can there be to enhance one's reputation? Even if some unfortunate accident occurred during the job, if the overall record is commendable, it is appropriate to express concern for the people who were affected by the accident and explain what has been done to avoid similar incidents in the future. If the person who was injured can appear at the ceremony and be congratulated on his recovery, some measure of good will may be reclaimed and some of the unfavorable publicity that may have followed the accident could perhaps be neutralized. Just what should be done at the end of a job cannot be prescribed in advance, and certainly I have no crystal ball that will provide the answers. However, I have often noted opportunities that were missed because someone was too preoccupied with the next day's work to take advantage of the chance to participate in a celebration of a completed job of which he should be proud. How better can one show one's pride in an accomplishment than to congratulate and praise those who helped make it possible? No amount of advertising or self-praise can equal the good will of people whose worth has been recognized publicly. Safety programs involve a lot of hard work; they should include as much fun as possible whenever there is an excuse for a party and a happy story to be told.

APPENDIX A
OSHA Recordkeeping and Incidence Rates

OSHA regulations 29CFR Part 1904 require that every accident resulting in one or more deaths or in hospitalization of five or more employees be reported to the OSHA area director within 48 hours. (See Appendix B for a list of locations.) The report may be verbal or in writing. I suggest that it be made verbally as soon as possible after any accident that will probably require action and confirmed after further details are known. This will give OSHA the opportunity to investigate more promptly. Be sure to record the date, time, and person's name to whom any verbal report is made, and include those details in the written confirmation. If a fatality is reported and it is later discovered that the person survived, it is still better to have made the earliest possible report than to have waited for further developments.

A second type of record required by OSHA is illustrated by reference to 29CFR 1926.850 on demolition. The opening paragraph states that an engineering survey must be made before work begins and the employer must have written evidence that the survey was made. No time limit is set for retaining the record, and presumably it may be discarded when the project is complete unless claims for damages are still pending. Reports need not be sent to OSHA unless requested.

A third and somewhat similar type of record requirement is mandated by 29CFR 1910.1001 on asbestos dust. Records may have to be kept for well over 20 years, but need not be sent to OSHA or disclosed to workers or their representatives unless requested. Such rules are difficult to apply to construction; exposure is difficult to monitor over a full work day, and is apt to be highly variable from one day to the next. If compliance with any regulation of this nature is required, I suggest that the safety engineer request a current interpretation from the OSHA regional office. Any comments I might make today could easily be obsolete tomorrow.

Every year the Bureau of Labor Statistics collects data on construction industry injuries and occupational illnesses. Not all employers are required to submit data; wait for a request before reporting.

All employers must keep records of "recordable" accidents and illnesses with the exception of firms with no more than 10 employees, and every year OSHA requests that a representative sampling of such small companies also maintain the records. These records must be kept for five years after the close of the current year.

The first of two such records is OSHA form 200, the "Bureau of Labor Statistics Log and Summary of Occupational Injuries and Illnesses." From February 1 to March 1 of each year, post a signed copy showing summary information at every location where employees are still working. Keep the complete logs and employment records for the current and the five preceding years; they must be available for inspection and copying by representatives of the government, employees, former employees, and their representatives. A separate log must generally be kept for each site, but it is permissible to hold them all at one location until it is time to post the summaries.

Figure A.1 shows the general method of telling whether or not an injury or illness is "recordable." It is reproduced from U.S. Department of Labor, Bureau of Labor Statistics, Report 412-3, "What Every Employer Needs to Know About OSHA Recordkeeping." Another useful booklet "Recordkeeping Requirements Under the Occupational Safety and Health Act of 1970" is published by OSHA. It includes sample forms and instructions for filling them out. Both publications are available from offices listed in Appendix B.

The person responsible for maintaining the log is allowed six days to enter the essential information after he learns that a case should be recorded. Revisions to the entry may be made at any time until a case is closed, even though a year or more may have elapsed since the initial entry. I have found it best to log every case immediately, even before being sure it is recordable. Inappropriate entries may be lined out, but having them all in the log makes it easier to evaluate the safety program at the end of the year and avoids the possibility that a recordable case may be overlooked. In fact, I think form 200 is one of OSHA's most useful tools for the safety engineer; even small contractors who are not required to maintain a log may find it an easy way to keep a record of pending claims.

A word of caution: never log the same case twice. If it is not closed by the end of the year in which the accident occurred, make the entry by using the best information available at the moment, and revise it later if necessary. Some cases drag on for years, but they appear only on the report for the year in which they occurred.

There may occasionally be difficulty telling whether a case is recordable. The employer makes the final decision, but he may ask OSHA, a doctor, or anyone else for advice.

The number of lost workdays may also be hard to determine. My advice is to follow the rules and not to try to use common sense when the regulations seem arbitrary or unreasonable. For example, if an injury occurs on a Friday and the worker was scheduled to be on vacation for the following two weeks, no workdays would be lost if he returns to the job at the end of his vacation. The day on which an injury occurs (Friday in this

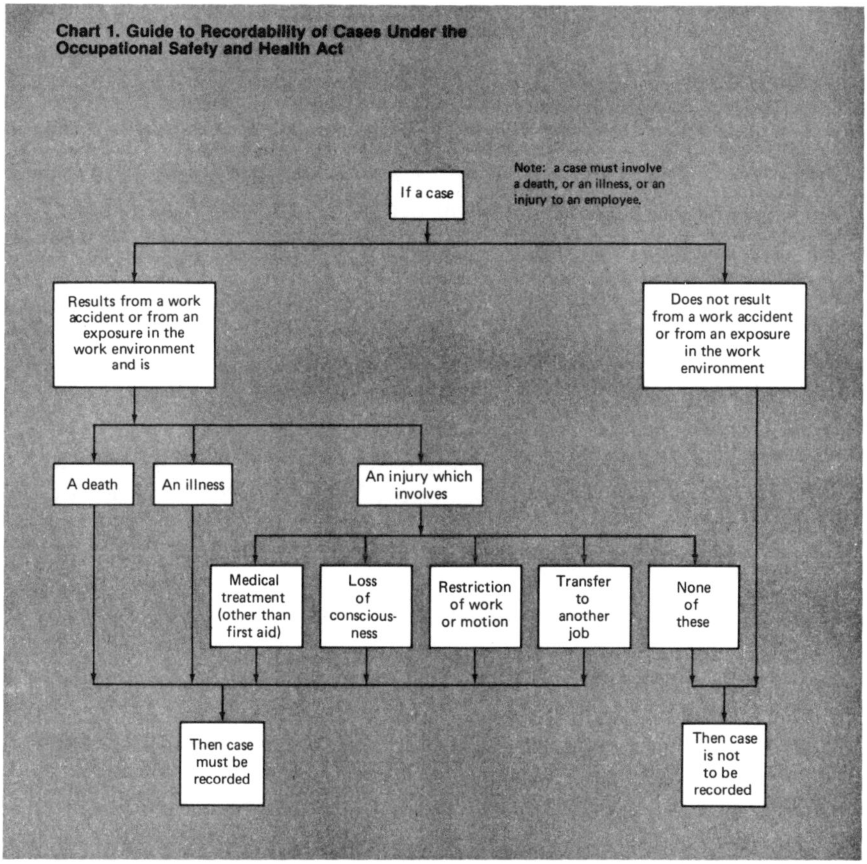

Figure A.1 How to determine whether an accidental death, injury, or illness is recordable under OSHA rules.

example) is never counted as a lost workday. On the other hand, if the same injury had occurred on any other workday and the employee could not report to the job the following day, it would be a lost workday case. The publications previously mentioned explain the rules for determining lost workdays and OSHA will help interpret the regulations if asked.

The second record that all employers of over 10 people must maintain is OSHA Form 101 "Supplementary Record of Occupational Injuries and Illnesses" *or its equivalent.* Get a copy from one of the offices in Appendix B and compare it carefully with the claim form used by your insurance company. OSHA does not care what form is used, provided all information requested on their form 101 is available. Do not overlook the case number. Time may be saved by using the insurance form, even if a photocopy is made for the OSHA records. Discrepancies between reports are avoided when this routine is chosen.

Most safety engineers like to know whether they have accomplished anything during the year and how their company's safety record compares with others. Logs of previous years are interesting, but incidence rate calculations provide additional insight. They are readily calculated, provided that employment records are correlated with the log sheets to indicate the number of man-hours worked at the location covered by each log. Incidence records are calculated on the basis of 200,000 man-hours per year, equivalent to the number of hours worked by 100 people, 40 hours per week, 50 weeks per year.

Consider first the incidence rate for total recordable cases:

$$\text{Incidence rate} = \frac{\text{Number of recordable cases} \times 200{,}000}{\text{Employee hours worked}}$$

For example, assume you are doing nonresidential construction and had 15 recordable cases during the past year. Your time records indicate that employees worked a total of 157,000 hours. The incidence rate would be:

$$\frac{15 \times 200{,}000}{157{,}000} = 19.1$$

This rate can be compared with your own for previous years, and with statistics published by the Bureau of Labor Statistics. Table A.1 is the first section of a longer table in the Bureau's Report, "Evaluating Your Firm's Injury and Illness Rate, 1977."

In the example, 157,000 man-hours is equivalent to having about 78 or 79 people working 40 hours per week for 50 weeks during the year, and therefore figures in the fourth line of Table A.1 should be inspected. Your rate of 19.1 is somewhat better than the group average of 20.7 given in column A of the table. On the other hand, column B shows that one quarter of comparable firms had incidence rates below 10.0 and this could be an appropriate goal for you to use for the coming year. Column C of the table shows the median incidence rate was 20.1, Column D shows that the quarter of the companies with the worst accident records had incidence rates of 29.8 or above; almost one-third of their workers were injured during the year.

Safety engineers will, of course, immediately recognize that statistics based only on the total number of cases can tell only part of the story. One of two companies might have 10 fatalities and another of comparable size might have 10 minor recordable accidents, but both would have the same incidence rates. Some measurement of severity is obviously required and several formulas are in common use, similar in form to the one already discussed, but taking into account the number of fatalities, lost workdays, transfers, and light duty assignments. I cannot devote enough space to this topic to even begin to treat it adequately, but anyone who wants to pursue

TABLE A.1 Occupational Injury and Illness Incidence Rates for Nonresidential General Building Contractors by Employer Size and Quartile Distribution, United States, 1977

	Incidence Rates per 100 Full-Time workers			
	Column A	Column B	Column C	Column D
Industry and Employment Size	Average Incidence Rates for All Establishments: (mean)	One-Quarter of the Establishments Had a Rate Lower Than or Equal to: (1st quartile)	One-half of the Establishments Had a Rate Lower Than or Equal to: (median)	One-Quarter of the Establishments Had a Rate Higher Than or Equal to: (3rd quartile)
General building contractors				
Nonresidential building construction:				
All sizes	17.9	0.0	0.0	0.0
1 to 19	12.0	0.0	0.0	15.5
20 to 49	18.9	5.1	15.7	29.0
50 to 99	20.7	10.0	20.1	29.8
100 to 249	22.2	9.2	20.7	33.5
250 to 499	20.0	9.4	19.0	32.9
500 to 999	19.9	10.7	16.6	27.2

it further will be interested in the American National Standard for Uniform Recordkeeping for Occupational Injuries and Illnesses, ANSI Z 16.4-1977. Copies are available from the American National Standards Institute, 1430 Broadway, New York, N.Y. 10018. The National Safety Council is another good source of information at 444 North Michigan Avenue, Chicago, Illinois 60611.

But safety supervisors should beware of focusing their attention exclusively on recordable cases. Minor accidents and near misses can be equally significant indications of hazards. Only a few months ago a crane tipped over at a job in Stamford, Connecticut; fortunately nobody was hurt, but while I am writing this a fatal crane accident at the same site is being investigated. Perhaps if the first accident had led to tighter controls on crane operations the second one might have been averted.

APPENDIX B
OSHA and Bureau of Labor Statistics Regional Offices

Regional offices of the U.S. Department of Labor, Occupational Safety and Health Administration, are at the following locations:

REGION I (CT, ME, MA, NH, RI, VT)
16-18 North Street
1 Dock Square, 4th Floor
Boston, MA 02109
Telephone: (617) 223-6710

REGION II (NY, NJ, PR, VI)
Room 3445, 1 Astor Plaza
1515 Broadway
New York, NY 10036
Telephone: (212) 944-3426

REGION III (DE, DC, MD, PA, VA, WV)
Gateway Bldg., Suite 2100
3535 Market Street
Philadelphia, PA19104
Telephone: (215) 596-1201

REGION IV (AL, FL, GA, KY, MS, NC, SC, TN)
1375 Peachtree Street, N.E.
Suite 587
Atlanta, GA 30367
Telephone: (404) 881-3573

REGION V (IL, IN, MI, MN, OH, WI)
230 South Dearborn Street
32nd Floor, Room 3230
Chicago, IL 60604
Telephone: (312) 353-2220

REGION VI (AR, LA, NM, OK, TX)
555 Griffin Square, Room 602
Dallas, TX 75202
Telephone: (214) 767-4731

REGION VII (IA, KS, MO, NE)
911 Walnut Street, Room 3000
Kansas City, MO 64106
Telephone: (816) 374-5861

REGION VIII (CO, MT, ND, SD, UT, WY)
Federal Bldg., Room 1554
1961 South Street
Denver, CO 80294
Telephone: (303) 837-5285

REGION IX (CA, AZ, NV, HI, Guam, American Samoa, Trust Territory of the Pacific Islands)
450 Golden Gate Avenue—Box 36017
San Francisco, CA 94102
Telephone: (415) 556-0584

REGION X (AK, ID, OR, WA)
Federal Office Bldg., Room 6003
909 First Avenue
Seattle, WA 98174
Telephone: (206) 442-5930

Regional offices of the Bureau of Labor Statistics are at the following locations:

REGION I (BOSTON)
1603-A Federal Office Building
Boston, Massachusetts 02203
Telephone: (617) 223-4533
Connecticut
Maine
Massachusetts
New Hampshire
Rhode Island
Vermont

REGION II (NEW YORK)
1515 Broadway
New York, New York 10036
Telephone: (212) 944-3149
New Jersey
New York
Puerto Rico
Virgin Islands

REGION III (PHILADELPHIA)
Post Office Box 13309
Philadelphia, Pennsylvania 19101
Telephone: (215) 596-1162
Delaware
District of Columbia
Maryland
Pennsylvania
Virginia
West Virginia

REGION IV (ATLANTA)
1371 Peachtree Street, N.E.
Atlanta, Georgia 30309
Telephone: (404) 881-3660

Alabama	Mississippi
Florida	North Carolina
Georgia	South Carolina
Kentucky	Tennessee

REGION V (CHICAGO)
Post Office Box 2145
Chicago, Illinois 60690
Attn: OSHA Forms
Telephone: (312) 353-1880
Illinois
Indiana
Michigan
Minnesota
Ohio
Wisconsin

REGION VI (DALLAS)
555 Griffin Square Building
2nd Floor
Dallas, Texas 75202
Telephone: (214) 767-6954
Arkansas
Louisiana
New Mexico
Oklahoma
Texas

REGIONS VII AND VIII
(KANSAS CITY AND DENVER)
Federal Office Building
911 Walnut Street
Kansas City, Missouri 64106
Telephone: (816) 374-3685

Colorado	Nebraska
Iowa	North Dakota
Kansas	South Dakota
Missouri	Utah
Montana	Wyoming

REGIONS IX AND X
(SAN FRANCISCO AND SEATTLE)
450 Golden Gate Avenue
Box 36017
San Francisco, California 94102
Telephone: (415) 556-8980

Alaska	Hawaii
American Samoa	Idaho
Arizona	Nevada
California	Oregon
Guam	Washington

APPENDIX C
Some Dangerous Substances

Every substance is harmful in excess. Even such essentials as oxygen and water can cause illness or death if administered in excess of the body's normal requirements. Many people have an irrational fear of chemicals, forgetting that the term describes everything in nature as well as manmade compounds. Moreover, humans have a considerable tolerance for thousands of chemicals that are used as food additives, medicines, cosmetics, and just about everything we buy in the course of a typical trip to the supermarket. We tend to overlook the importance of the concentrations of the chemicals. They are dangerous only if present in excess of the body's tolerance. There is no known material for which we have absolutely zero permissible exposure level. Sometimes the permissible concentration is measured in percent of the atmosphere, and sometimes it is smaller and is measured in parts per million; highly toxic substances may have to be limited to concentrations measured in parts per billion. Unfortunately, politicians and laypersons often fail to realize the great difference between 1 part per million and one part per billion. They think that setting limits as low as possible is best for the public safety and as a result wastes may have to be treated to a purity far greater than that of the streams into which they are discharged. Money squandered on this type of overtreatment is no longer available for more realistic waste treatment activities, and the net result on the environment is just the opposite of the intent of the legislators or ill-informed environmentalists. Attention should be given to actual dangers, based on scientific findings.

Thousands of chemicals may be considered dangerous; one list of 60,000 items was prepared by the National Institute for Occupational Safety and Health. Naturally there is controversy about which items should be listed and the extent to which they are hazardous. OSHA general industry regulations from 29CFR 1910 sometimes apply to construction work, and those listed in Table C.1 have been designated as part of 29CFR 1926, which governs construction work. The sections on asbestos, coal tar pitch volatiles, vinyl chloride, and benzene are recommended reading for all safety engineers. Other references should be studied when the products listed appear on labels of materials to be used.

I want to comment on only a few substances that may be of interest to contractors because they are typical of certain groups of products used in construction or illustrate common dangers.

TABLE C.1

29CFR Paragraph Number	Toxic or Hazardous Substance
1910.1001	Asbestos
1910.1002	Coal tar pitch volatiles; interpretation of term
1910.1003	4-Nitrobiphenyl
1910.1004	alpha-Naphthylamine
1910.1005	4, 4'-Methylene bis(2-chloroaniline)
1910.1006	Methyl chloromethyl ether
1910.1007	3, 3'-Dichlorobenzine (and its salts)
1910.1008	bis-Chloromethyl ether
1910.1009	beta-Naphthylamine
1910.1010	Benzidine.
1910.1011	4-Aminodiphenyl
1910.1012	Ethyleneimine
1910.1013	beta-Propiolactone
1910.1014	2-Acetylaminofluorene
1910.1015	4-Dimethylaminoazobenzine
1910.1016	N-Nitrosodimethylamine
1910.1017	Vinyl Chloride
1910.1018	Inorganic arsenic
1910.1028	Benzene
1910.1029	Coke Oven Emissions
1910.1043	Cotton dust
1910.1044	1,2-dibromo-3-chloropropane
1910.1045	Acrylonitrile
1910.1046	Exposure to cotton dust in cotton gins

Carbon dioxide could hardly be classified as a highly toxic substance, but under some circumstances it can cause fatalities. It is, of course, the gas that forms the bubbles in champagne and soft drinks, it is sold in solid form as dry ice, and is used in fire extinguishers for fires on which water may not be sprayed. It is naturally present in the atmosphere, is formed in the body by the oxidation of carbon in foods, and exhaled with every breath. Yet, if the percentage of carbon dioxide in the air is modified by only a few percent, it will affect blood pressure, pulse rate, and hearing. It can cause falls and death by asphixiation.

Nitrogen is another gas that is not usually considered dangerous; it makes up about 78% of the atmosphere. Still, if enough compressed nitrogen is released in an enclosed space, it can reduce the percentage of oxygen in the air and cause asphixiation. Nitrogen compounds are among the most dangerous chemicals. Welding operations cause some of the oxygen and nitrogen in the air to combine, forming nitrogen dioxide and other nitrogen oxides that are very toxic. In confined spaces where welding fumes are not properly vented, dangerous concentrations may form. Most explosives are nitrogen compounds, and both ammonia and nitric acid are compounds of nitrogen and hydrogen.

Carbon monoxide poisoning is so often attributed to engine exhaust fumes that many people think it has a distinctive odor whereas in fact it is completely odorless. Incomplete combustion of charcoal or methane can produce carbon monoxide with practically no accompanying odor. In case of fire, many plastics decompose and release carbon monoxide. This gas interferes with the ability of the red blood cells to carry oxygen and, if the first warning signs (dizziness, ringing of the ears, pounding of the heart) are not heeded, death may quickly follow. Respirators that use absorbent cartridges cannot be used safely even if they can absorb the gas because the user has no warning when the cartridge has become exhausted.

Refrigerants, such as R12, R21, R114, etc., have very little toxic effect when small quantities are inhaled but, if the liquefied gas or the rapidly expanding vapor comes in contact with the skin, it may freeze tissues and cause irreversible damage. If a pipe or flexible connection breaks, releasing the whole charge in a large airconditioning system into a confined space, the percentage of oxygen in the air may be dangerously reduced by displacement. If any of these refrigerants comes in contact with flames or hot metal, they will decompose and release highly toxic gases.

Halogenated hydrocarbons constitute a large and important group of industrial chemicals, including carbon tetrachloride, methyl chloroform, and vinyl chloride. The vapors of some of these compounds can cause a person to lose consciousness, suffer heart or respiratory failure, or the kidneys, liver, and nerves may be damaged. If they are spilled on clothing or gloves are wet with them, some halogenated hydrocarbon compounds are capable of penetrating the skin and most will be hazardous when inhaled as vapors. Most people over 50 will remember carbon tetrachloride as a popular spot remover and filler for fire extinguishers, but both these applications have been banned in most communities because the vapors are toxic and the liquid decomposes on contact with flames to release phosgene. Watch labels; many products still contain carbon tet.

Methyl bromide was recently used to fumigate two warehouses in New York City that were infested with khapra beetles. The news media had a field day because the buildings had to be totally enshrouded with vapor proof tarps, streets were closed off, and people were excluded from adjacent buildings. Protesting residents demonstrated in the streets and there were false reports that birds and pets had died. When the job was finally completed all the reports of injuries were found to be ungrounded, but the dangers of using methyl bromide had been well publicized. No live beetles were found after the fumigation.

Hydrogen sulfide and rotten eggs smell alike, but since the development of home refrigerators many people know the odor of the gas who have never encountered an offensive egg. A more common source of the gas is bacterial decomposition of sulfur-bearing organic wastes in sewers or land fill. Breathing a small quantity of hydrogen sulfide can make a person temporarily unable to detect the odor. Then, if there is exposure to a high concentration of the gas the individual may lose consciousness and die. It can happen very fast, so if a worker notes nausea, eye irritation, dizziness, or a headache in a place where hydrogen sulfide might be expected to be present, he should get out of the contaminated area without delay. High concentrations of the gas in air are flammable and explosive.

Alcohols are popular solvents, thinners, and antifreeze compounds. Most distilled beverages are marketed at 80 to 100 proof, indicating they contain 40 to 50% ethanol or ethyl alcohol. Industrial grades are likely to be sold at much higher proof and usually contain additives to keep users from converting them to beverages. Vapors of strong ethanol can cause dizziness, headache, irritation of nasal passages, and loss of consciousness.

During prohibition, methanol (methyl alcohol or wood alcohol) was often substituted for grain alcohol, ethanol, and either death or blindness was frequently the result. Liquid and vapors of methanol can cause poisoning and skin irritation. Watch for it in fuel additives and solvents as well as antifreeze mixtures.

Ethylene chlorohydrin, or 2 chloroethanol is used as a solvent in lacquers. It is particularly dangerous because its odor is not irritating, but the vapor is capable of causing death or permanent injury after short exposure. The liquid will penetrate rubber gloves, and fatal amounts can be rapidly absorbed through the skin. By absorbing the liquid and holding it against the hands, rubber gloves may increase the danger to workers and at the same time give them a false sense of security. When it is heated to the decomposition temperature, this solvent releases phosgene. That gas, well known to be poisonous, may have a delayed reaction, causing death a day or two after exposure.

Even after the dangers associated with the use of gasoline have been publicized for generations, accidents continue to happen, particularly when spills occur and when it is used to clean oily materials from the hands or equipment. Possibly workers are ignorant of the high density of gasoline vapors, and do not realize that ignition with explosive violence can occur beyond the visible limits of a spill. At gasoline stations, hoses are grounded to prevent static electricity from causing sparks that can ignite vapors. Workers who pour gasoline from cans are not automatically protected from this hazard. Familiarity breeds contempt, and it is hard to keep workers alert to the dangerous properties of a fuel that is so commonly used.

Hydrochloric acid (muriatic acid) is commonly sold in dilute form and, when it is used to clean masonry or to descale equipment, the fumes of hydrogen chloride are sufficiently irritating to make workers glad to wear respirators. When hydrochloric acid is to be used to clean metal, it should always include an inhibitor to minimize corrosion and the release of hydrogen gas, which is explosive. Lightweight polyethylene containers with a corrugated overpack may be used to ship this acid, but after prolonged storage vapor may penetrate the plastic and deteriorate the overpack. If this happens, a pile of containers may collapse or a worker may drop one and be splashed. Use up the acid as soon as possible after it is received.

Hydrochloric acid in the form of a gas is one of the most dangerous ingredients of smoke. It is released when many plastics, including polyvinyl chloride (PVC) and many others are decomposed by heat. It is an irritating and choking gas, capable of causing eye and lung damage; it is a common cause of fatalities.

Sulfuric acid has a well-deserved bad reputation, but workers are less likely to have had direct experience with it than with muriatic acid. The concentrated acid causes severe burns and the fumes are irritating; workers need to be well trained before being allowed to handle it, and they need adequate protective gear. Sulfuric acid is often the proper material to add to condenser cooling water to control scale formation and, when operating engineers have the right training and equipment, there is no reason for them to be injured. However, construction workers may be expected to apply treatment during the initial operating period, and they may not do the job safely. I suggest that contractors investigate the possibility that use of sulfuric acid may be deferred until the regular operating staff takes over. Sometimes sulfamic acid, which costs more but is a granular solid, may do the job, particularly if only a small amount of acid is needed per day.

Workers may be surprised to find that concentrated sulfuric acid is shipped in plain steel drums. It is not corrosive when highly concentrated, but if the bung of a drum is left open, the acid will draw moisture from the air and a dilute layer of acid at the surface will become very corrosive to the steel. Hydrogen gas will be released. Consequently, the drum may leak, or if a source of ignition comes near the bung, the hydrogen gas inside the drum may explode. If water is added to concentrated sulfuric acid, heat is released and the mixture may be spattered onto the worker. *Always pour acid into an excess of water; never pour water into the acid.*

Watch labels on products sold for clearing clogged drains; they may be composed of sulfuric acid and little else. Other drain cleaners may be composed of strong caustics, such as sodium hydroxide or lye. In some respects the strong caustics are more dangerous than the strong acids; they can be difficult to wash off the skin and they saponify fatty tissues. Caustic soda (sodium hydroxide) granules will draw moisture from the air and form little drops of solution where ever they happen to fall. The solution can destroy the soles of leather shoes. Severe injury followed spillage of dry powdered caustic into a worker's shoe. The initial burn was severe enough to numb his nerves, and he did not realize that further damage was being done after the first treatment by some chemical that remained in the shoe. Water poured onto dry caustic or into strong liquid caustic will cause rapid heat evolution and dangerous spattering is to be expected.

If the powder is poured onto a small amount of water, steam may be produced fast enough to disperse the material through the air. To dissolve granular caustic, always pour it slowly into an excess of water and stir until each addition has dissolved before adding more. Don't rush; boiling hot caustic is one material to be avoided at all costs.

The caustic group of chemicals includes sodium hydroxide, potassium hydroxide, and trisodium phosphate, which are powders. Liquid caustic soda and various grades of sodium silicate are also common ingredients of strong cleaning products. The dusts are irritating to the eyes and respiratory tract and, if liquid sodium silicate (water glass) dries on any flexible surface and is dislodged, the sharp glasslike particles can cause painful cuts that are likely to be slow to heal.

Wood preservatives often contain pentachlorophenate dissolved in mineral spirits. The chemical can be absorbed through the skin and prolonged exposure can damage the kidneys and liver. The same active chemical in briquette form is sometimes added to condenser cooling water to kill organic growths in cooling towers or condenser tubes. The dry material has a strong odor and the dust is poisonous. Inhalation of even a small amount will probably cause prolonged discomfort. In general all biocides must be handled with care.

When chemicals are stored on the job, it is essential that piles of materials that would react with each other in case a spill occurred be kept well separated, but it is not always easy to tell which chemicals are dangerous when mixed. Even materials that carry the same Department of Transportation hazardous material label can react violently with each other. For example, sulfuric acid and sodium hydroxide are both "Corrosive Materials," but it would be hard to find two common chemicals that would be more dangerous to store together. Only a person who is very familiar with all the properties of every material that is to be stored should attempt to select those that are compatable. The safe procedure is to keep every one in a separate pile and keep the piles as far apart as possible. Flammable liquids should be kept out of doors or in fire-resistant cabinets when approved by local authorities.

I have touched only on the properties of a few common materials, but the need to take precautions is clear. In conclusion, I have but two recommendations:

1. Read every label carefully.
2. Take it seriously. Some may exaggerate dangers but many do not.

APPENDIX D
Shade Numbers for Welder's Goggles or Face Shields

The following table is reproduced from OSHA regulations, 29CFR 1926.102. Shades more dense than those listed may be used to suit the individual welder's needs.

Filter Lens Shade Numbers for Protection Against Radiant Energy

Welding Operation	Shade Number
Shielded metal-arc welding 1/16, 3/32, 1/8, 5/32 inch diameter electrodes	10
Gas-shielded arc welding (nonferrous) 1/16, 3/32, 1/8, 5/32 inch diameter electrodes	11
Gas-shielded arc welding (ferrous) 1/16, 3/32, 1/8, 5/32 inch diameter electrodes	12
Shielded metal-arc welding 3/16, 7/32, 1/4 inch diameter electrodes	12
5/16, 3/8 inch diameter electrodes	14
Atomic hydrogen welding	10–14
Carbon-arc welding	14
Soldering	2
Torch brazing	3 or 4
Light cutting, up to 1 inch	3 or 4
Medium cutting, 1 inch to 6 inches	4 or 5
Heavy cutting, over 6 inches	5 or 6
Gas welding (light), up to 1/8 inch	4 or 5
Gas welding (medium), 1/8 inch to 1/2 inch	5 or 6
Gas welding (heavy), over 1/2 inch	6 or 8

APPENDIX E
Electric Extension Cord Capacities

Always use extension cords that have adequate capacity. A cord that is too small has excessive resistance, causing low voltage at the load. This reduces the available power and can permanently damage portable tools. It can also cause the cord to overheat and in some cases cord resistance may be so high that it will prevent a fuse from blowing when there is a defect in the tool or the cord insulation is damaged. This can cause a fire. Weatherproof three conductor grounding type cords with oilproof insulation are recommended for construction jobs. The following table was supplied by Skil Corporation, 4801 W. Peterson Avenue, Chicago, Illinois 60646. See Section 7.3 for comments on small tool safety.

Recommended Minimum Gage
For Cord Extension for Portable Electric Tools

Ampere Rating Range	Volts 120v 240v	26 ft 50 ft	50 ft 100 ft	100 ft 200 ft	150 ft 300 ft	200 ft 400 ft	250 ft 500 ft	300 ft 600 ft
0-2		18	18	18	16	16	14	14
2-3		18	18	16	14	14	12	12
3-4		18	18	16	14	12	12	10
4-5		18	18	14	12	12	10	10
5-6		18	16	14	12	10	10	8
6-8		18	16	12	10	10	8	6
8-10		18	14	12	10	8	8	6
10-12		16	14	10	8	8	6	6
12-14		16	12	10	8	6	6	6
14-16		16	12	10	8	6	6	4
16-18		14	12	8	8	6	4	4
18-20		14	12	8	6	6	4	4

Source: Skill Corporation

APPENDIX F
Illumination Required in Work Areas

OSHA regulations, 29CFR 1926.56 require work areas to be lighted to not less than the intensities listed in the following table while any work is in progress. For areas not listed, American National Standard A 11.1-1965 R1970 is applicable.

Minimum Illumination Intensities in Foot-Candles

Foot-Candles:	Area or Operation
5	General construction area lighting.
3	General construction areas, concrete placement, excavation and waste areas, accessways, active storage areas, loading platforms, refueling, and field maintenance areas.
5	Indoors: warehouses, corridors, hallways, and exitways.
5	Tunnels, shafts, and general underground work areas: (Exception: minimum of 10 foot-candles is required at tunnel and shaft heading during drilling, mucking, and scaling. Bureau of Mines approved cap lights shall be acceptable for use in the tunnel heading.)
10	General construction plant and shops (e.g., batch plants, screening plants, mechanical and electrical equipment rooms, carpenter shops, rigging lofts and active storerooms, barracks or living quarters, locker or dressing rooms, mess halls, and indoor toilets and workrooms).
30	First aid stations, infirmaries, and offices.

APPENDIX G
Respirator Selection For Various Hazards

The following table appears in 29CFR 1926.103, and is a summary of respirator types that are suitable for many job conditions. Observe that the same types appear repeatedly in the table. Before buying any one type, check all possible job requirements to see whether a more versatile style would cover all needs and thereby eliminate the possibility that the proper one might not be at hand in an emergency. See Section 3.5.3 for descriptions of respirators and Section 7.7 for comments on enclosed space-hazards.

Selection of Respirators

Hazard	Respirator (See Note)
Oxygen deficiency	Self-contained breathing apparatus. Hose mask with blower. Combination air-line respirator with auxiliary self-contained air supply or an air-storage receiver with alarm.
Gas and vapor contaminants immediately dangerous to life and health.	Self-contained breathing apparatus. Hose mask with blower. Air-purifying, full facepiece respirator with chemical canister (gas mask). Self-rescue mouthpiece respirator (for escape only). Combination air-line respirator with auxiliary self-contained air supply or an air-storage receiver with alarm.
Not immediately dangerous to life and health	Air-line respirator. Hose mask without blower. Air-purifying, half-mask or mouthpiece respirator with chemical cartridge.
Particulate contaminants immediately dangerous to life and health	Self-contained breathing apparatus. Hose mask with blower. Air purifying, full facepiece respirator with appropriate filter. Self-rescue mouthpiece respirator (for escape only). Combination air-line respirator with auxiliary self-contained air supply or an air-storage receiver with alarm.

Not immediately dangerous to life and health	Air-purifying, half-mask or mouthpiece respirator with filter pad or cartridge. Air-line respirator. Air-line abrasive-blasting respirator. Hose mask without blower.
Combination gas, vapor, and particulate contaminants immediately dangerous to life and health.	Self-contained breathing apparatus. Hose mask with blower. Air-purifying, full facepiece respirator with chemical canister and appropriate filter (gas mask with filter). Self-rescue mouthpiece respirator (for escape only). Combination air-line respirator with auxiliary self-contained air supply or an air-storage receiver with alarm.
Not immediately dangerous to life and health	Air-line respirator. Hose mask without blower. Air-purifying, half-mask or mouthpiece respirator with chemical cartridge and appropriate filter.

Note: For the purpose of this part, "immediately dangerous to life and health" is defined as a condition that either poses an immediate threat to life and health or an immediate threat of severe exposure to contaminants, such as radioactive materials, which are likely to have adverse delayed effects on health.

APPENDIX H
Scaffolds

For complete coverage of OSHA regulations governing all types of scaffolds and elevated work platforms, see 29CFR 1926.451. In it, tables L4 to L9 deal with wood pole and metal pole and coupler scaffolds that are commonly built from lumber or metal tubing to suit local requirements. To erect them, workers need to know the proper sizes of lumber or tubing to use and the corresponding limitations on loading and height. I found the OSHA tables repetitive, and have combined them into Tables H.1, H.2, and H.3 in the hope that this would make the essential information easier to find.

One of the standardized tubular welded steel assemblies will often be appropriate if the height does not exceed 125 feet, but for all scaffolds over that height be sure that the design was prepared by a registered professional engineer.

I have not attempted to present a summary of rules applicable to all the other types of scaffolds; safety engineers should refer directly to the regulations in 29CFR451.

TABLE H.1 Single Pole Wood Scaffolds

	Light Duty		Medium Duty	Heavy Duty
Maximum height, feet	20	60	60	60
Maximum uniformly distributed load, pounds/square foot	25	25	50	75
Poles or uprights	2×4	4×4	4×4	4×6
Pole spacing, longitudinal	6′0″	10′0″	8′0″	6′0″
Maximum width of scaffold	5′0″	5′0″	5′0″	5′0″
Bearers or putlogs[a]				
To 3′0″	2×4	2×4	2×10 or 3×4	2×10 or 3×4
To 5′0″ width	2×6 or 3×4 rough	2×6 or 3×4 rough	2×10 or 3×4	2×10 or 3×5
Spacing of bearers or putlogs	6′0″	10′0″	8′0″	6′0″
Ledgers	1×4	1¼×9	2×10	2×10
Planking	1¼×9 rough	2×10	2×10	2×10

TABLE H.1 Continued

	Light Duty		Medium Duty	Heavy Duty
Vertical spacing of horizontal members	7′0″	9′0″	7′0″	6′6″
Bracing, horizontal	1×4	1×4	1×6 or 1¼×4	2×4
Bracing, diagonal	1×4	1×4	1×4	2×4
Tie-ins	1×4	1×4	1×4	1×4
Standard guardrails	[b]	[b]	[b]	[b]
Reference: table number in 29CFR 1926.451	L4	L4	L5	L6

[a]Putlog must be reenforced with 3/16″ × 2″ steel strip or equivalent secured to its lower edge along its entire length.
[b]See Chapter 6, Section 6.3.1 regarding guardrails; required if scaffold is over 10′ high. Scaffolds 4′ to 10′ high and less than 45″ wide or long also require guardrails.

TABLE H.2 Independent Pole Wood Scaffolds

	Light Duty		Medium Duty	Heavy Duty
Maximum height, feet	20	60	60	60
Maximum uniformly distributed load, pounds/square foot	25	25	50	75
Poles or uprights	2×4	4×4	4×4	4×4
Pole spacing, longitudinal	6′0″	10′0″	8′0″	6′0″
Pole spacing, transverse	6′0″	10′0″	8′0″	8′0″
Ledgers	1¼×4	1¼×9	2×10	2×10
Bearers				
To 3′0″ span	2×4	2×4	—	—
To 10′0″ span	2×6 or 3×4	2×10 rough or 3×8	2×10	2×10
Spacing of bearers	6′0″	10′0″	8′0″	6′0″
Planking	1¼×9	2×10	2×10	2×10
Vertical spacing of horizontal members	7′0″	7′0″	6′0″	6′0″
Bracing, horizontal	1×4	1×4	1×6 or 1¼×4	2×4
Bracing, diagonal	1×4	1×4	1×4	2×4
Tie-ins	1×4	1×4	1×4	1×4
Standard guardrails	[a]	[a]	[a]	[a]
Reference: table number in 29CFR 1926.451	L7	L7	L8	L9

[a]See Chapter 6, Section 6.3.1 regarding guardrails; they are required if scaffold is over 10′0″ high. Scaffolds 4′ to 10′ high and less than 45″ wide or long also require guardrails.

TABLE H.3 Tube and Coupler Scaffolds

		Light Duty	Medium Duty	Heavy Duty
Maximum heights, feet				
Working Levels	Additional Planked Levels			
1	8	125	—	—
2	4	125	—	—
3	0	91	—	—
1	6	—	125	—
2	0	—	78	—
1	6	—	—	125
Maximum uniformly distributed load, pounds per square foot		25	50	75
Tubing size, nominal O.D. size for posts, runners, and bracing		2″	2″	2″
Tubing size for bearers		2″	2½[a]	2″
Post spacing, longitudinal		10′0″	8′0″	6′6″
Post spacing, transverse		6′6″	6′6″	6′6″
Standard guardrails		[b]	[b]	[b]
Reference: table number in 29CFR 1926.451		L10	L11	L12

[a]For post spacing of 8′0″ longitudinal by 5′0″ transverse, 2″ O.D. tubing may be used for bearers

[b]See Chapter 6, Section 6.3.1 regarding guardrails; they are required if scaffold is over 10′ high. Scaffolds 4′ to 10′ high and less than 45″ wide or long also require guardrails.

APPENDIX I
Job-Built Cleat Type Ladders

OSHA Regulations 29CFR 1926.450(b) govern this type of ladder. If 25 or more workers must use a job-built ladder as the only means of access, it must be of the double cleat (three rail) type. Otherwise, the single cleat (two rail) design is acceptable.

Table I.1, reproduced from the OSHA reference, lists the kinds of wood that may be used. Table I.2 summarizes OSHA information on permissible lengths of ladders and rail requirements. If the maximum ladder lengths, measured from one working level to the other, are insufficient, an intermediate platform with standard railings is required. Unless rails are at least three feet longer than the ladder lengths indicated, grab holds are required at the top level (Figure I.1).

Cleats must be set 12 in. top-to-top and secured with three 10d nails or equivalent fasteners at each rail. They must be either notched ½ in. into the rails or separated by filler blocks. Any kind of wood may be used provided it has strength equivalent to class 1 wood strips measuring ¾ × 3 in. if up to 20 in. long or ¾ × 3¾ in. if up to 30 in. long between rails. When woods other than class 1 are used, the thickness or width of the cleats must be increased, but regulations leave it to the builder's judgment to select suitable dimensions.

Rails may be parallel or flared no more than ⅛ in. per foot of length. The space between rails must be no less than 15 in. and no more than 20 in. at the top of single cleat ladders; presumably the same restrictions apply to double cleat ladders. If rails are spliced, the connections must be as strong as the rest of the rails, but the method of achieving this is not specified. Be sure that ladders are securely fastened in place and are kept in good condition; someone should check every shift for possible damage.

TABLE I.1 Average Densities of Various Species of Wood for Use in Ladders

Species	Density ($lb/ft.^3$)	Species	Density ($lb/ft.^3$)
	Group 1		
White ash	41	Hard maple	42
Beech	48	Red maple	36
Birch	44	Red oak	43
Rock elm	43	White oak	46
Hickory	50	Pecan	46
Locust	47	Persimmon	50
	Group 2		
Douglas fir		Southern yellow pine	37
(coast region)	34		
Western larch	38		
	Group 3		
Red alder	28	West Coast hemlock	30
Oregon ash	38	Magnolia	35
Pumpkin ash	37	Oregon maple	34
Alaska cedar	31	Norway pine	31
Port Orford cedar	30	Poplar	28
Cucumber	34	Redwood	25
Cypress	32	Eastern spruce	28
Soft elm	36	Sitka spruce	28
Douglas fir (Rocky		Sycamore	35
Mountain type)	30	Tamarack	37
Noble fir	27	Tupelo	35
Gum	34		
	Group 4		
Aspen	27	Eastern hemlock	28
Basswood	25	Holly	39
Buckeye	25	Soft maple	33
Butternut	27	Lodgepole pine	29
Incense cedar	25	Idaho white pine	28
Western red cedar	23	Northern white pine	25
Black cottonwood	24	Ponderosa pine	28
White fir	26	Sugar pine	26
Hackberry	37		

TABLE I.2 Maximum Allowable Lengths for Job-Built Cleat Type Ladders

Species of Wood Used for Rails (see Table I.1)	Nominal Size of Rails	Maximum Length between Base and Top Landing[a]	
		Single Cleat	Double Cleat
Group 2 or Group 3	2 × 4	16 ft	12 ft
	2 × 6	—	24 ft
	3 × 6	30 ft	—
Group 4	2 × 4	16 ft	12 ft
	2 × 6	—	20 ft
	3 × 6	20 ft	—

[a]Rails should extend at least 36 in. above top landing; otherwise secure grab holds must be provided at top.

Figure I.1 Both rails of this sturdy job-built ladder have been securely tied to the structure. Note blocking between cleats and observe that the rails extend about 4 or 5 feet above the top support. (Photo courtesy of Philadelphia Electric Company.)

APPENDIX J
*Sample of Confined Space Entry Permit**

Location of Work: ______________________________

Description of Work (Trades): ______________________________

Employees Assigned: ______________________________

Entry Date: ______________ Entry Time: ______________

Outside Contractors: ______________________________

Isolation Checklist:

- Blanking and/or Disconnecting
- Electrical
- Mechanical
- Other

Hazardous Work:

- Burning
- Welding
- Brazing
- Open Flame
- Other

Hazards Expected:

- Corrosive Materials
- Hot Equipment
- Flammable Materials
- Toxic Materials
- Drains Open
- Cleaning (Ex: chemical or water lance)
- Spark Producing Operations
- Spilled Liquids
- Pressure Systems
- Other

Vessel Cleaned:

- Deposits ______________________________
- Method ______________________________
- Inspection ______________________________
- Neutralized with ______________________________

Fire Safety Precautions: ______________________________

*See Chapter 7, Section 7.7.

Personal Safety:

Ventilation Requirements	Lighting
Respirators	Communications
Clothing	Employee Qualified
Head, Hand, and Foot Protection	Buddy System
Shields	Standby Person
Life Lines and Harness	Emergency Egress Procedures

Training Sign Off (Supervisor or Qualified Person) ________________

Remarks: ____

Atmospheric Gas Tests:

	Tests Performed –	Location –	Reading
Example:	(Oxygen)	______	(19.5%)
Example:	(Flammability)	______	(Less than 10% LFL)
	______	______	______
	______	______	______
	______	______	______

Remarks: ____

Test Performed By: ________________________________
Signature

Time: ____

Authorizations:

Supervisor: ________________________________

Prod Supervisor: ________________________________

Line Supervisor: ________________________________

Safety Supervisor: ________________________________

Etc.: ________________________________

Entry and Emergency Procedures Understood:

Standby Person ________________________________

Rescue ________________________________

Telephone ________________________________

Permit Expires: ________________________________

Classification: ________________________________

APPENDIX K
Checklist of Construction Safety Hazards

Every safety engineer, superintendent, and foreman should develop his own safety checklist and use it often enough to keep his awareness of safety hazards sharp at all times.

When several people participate in safety inspection tours, I have found it desirable for them to compare notes. No two men are likely to notice exactly the same items, and checklists are more useful if they are based on observations at the job as it progresses.

In the following checklist, the main topics are in alphabetical order. Unfortunately on any inspection tour the hazards do not line up like soldiers to be checked off. They appear in random order, and so it is impossible to arrange a checklist that will correspond to the needs of the field inspector in an entirely logical manner. I have found, however, that after using any list for a few days the principal items come to mind readily enough to make the alphabetical arrangement as practical as any other.

Blasting

1. Is the blaster fully qualified for the work to be done? Is he a competent person with full knowledge of the dangers involved? Does he have the authority to take responsible charge of the blasting work? Is he kept informed about the progress and scheduling of other work on the site?
2. Have other workers and residents of adjacent property been informed about the blasting so they will not be unnecessarily disturbed by it?
3. Are trained watchmen on duty when needed?
4. Are explosives and blasting equipment safe from theft?

Cranes

1. Who is responsible for crane inspections? Are they made and recorded? Are defects corrected before work proceeds?
2. Is the operator fully qualified? If a license is required, is his valid for the type and size crane he is operating?

3. How are lift weights determined? Does the operator know the permissible weights for each reach? Are automatic safe load indicators tested regularly?

4. Is the crane secure on a firm base and properly leveled?

5. Has the slinger been trained to give standard signals and to attach loads correctly?

6. Are knife-blade joints that may stick or jam assembled by lowering the load into position rather than by lifting it into engagement?

7. Are people kept out from under moving loads?

8. Are tag lines used?

9. Are critical operations discussed with all concerned before work begins, so that the reasons for all steps are understood? Has their proper sequence been agreed upon?

10. Are crane booms kept clear of power lines at all times?

11. Are slings inspected before each use? Is the capacity of each known? Does the slinger understand fully the effect of bridling? Can he estimate angles with sufficient accuracy to avoid excessive strain on slings?

Electricity

1. Are competent electricians available to install temporary wiring and to make any changes that are needed? Are untrained men prohibited from making changes or extensions?

2. Is a ground fault current interrupter system or an approved grounding system in use?

3. Is electrical equipment in good condition? Are extension cords regularly inspected and protected from damage in service? Are conductors large enough to carry the current without overheating?

4. Are all connections made by using appropriate plugs, receptacles, or enclosures? Are there any makeshift connections, bare wires, or damaged cables?

5. Have any machine operators reported tampering with the wiring of their equipment? Have vandals attempted to jump start or damage engines?

6. Are there any overhead electric power lines on the site? Are they where a crane might come close to them, or a workman might reach them from a scaffold or ladder? Are the lines disconnected when working close to them could be dangerous? What precautions are taken when it is necessary to work in the vicinity of hot lines?

7. If there is an underground cable on the site, has its exact location been confirmed? Is it always disconnected when work must be done that might cut into it?

8. Do workers habitually unplug electric tools before changing bits or moving them from one work area to another? Are hand lines always used to raise tools, rather than pulling them up by their electric cords?

9. Are metallic hard hats prohibited where contact with live wires might be possible?

Excavations

1. Who decides whether excavations should be shored? Is the cost of shoring likely to influence him to make an unwise decision?

2. How is the angle of slope of unshored excavations selected? Is the decision reviewed after the work has progressed far enough to expose different geological formations or soil conditions?

3. Is shoring inspected daily before work is resumed? Is it installed and removed in a safe manner? Is it adequately strong?

4. Is there safe access to the excavation and do means of escape comply with regulations? Are there any insecure ladders, planks, or ramps?

5. Is earth from the excavation piled well back from the edge? Would it be possible for any materials or rocks to roll or slide into the cut?

6. Is there a barrier to keep people and vehicles from falling into the excavation?

7. Is the stability of the excavation being disturbed by vehicles passing too close to it? If vehicles dump into the excavation, are their wheels blocked to prevent them from falling into the cut? Is there always someone watching while backfilling work is in progress?

8. If an unexpected pipe, tank, foundation, or other obstruction is found, is it investigated before work proceeds?

Fire

1. Are areas designated where smoking is permitted? Do workers refrain from smoking elsewhere?

2. Are fire extinguishers of appropriate size and type suitably located near hazardous work? Have workers been shown how to use them?

3. Is there a local fire department? In case of fire, is someone responsible for calling to report the fire immediately, even though it is anticipated that workers can get it under control themselves?

4. Do workers know how they are to proceed to a safe area? Do they know of alternative routes if the designated one is blocked?

5. Are highly flammable liquids stored in a properly located and constructed enclosure? Is the amount of flammable liquid available at the work area limited to no more than will be needed during the day, or to some other smaller amount designated by the fire marshal or by regulations?

6. Are approved safety containers used for flammable liquids and for contaminated rags or wipers?

7. Are compressed gases such as LPG and acetylene isolated from possible sources of ignition except when actually in use? Are cylinder valves kept closed unless gas is being used? Are LPG cylinders kept out of occupied shanties whenever possible?

8. Are flammable materials protected from sparks during welding or burning operations? Are watchers with fire extinguishers assigned when there is any possibility of fire?

9. Are oxygen and acetylene hoses protected from damage or kinking while in use? Are workers aware of the danger that a flashback could ignite the hose?

10. Are engine exhaust pipes kept clear of flammables? Are gasoline tanks and cans bonded together while gasoline is being transferred in order to avoid the possibility that vapors may be ignited by static electric sparks?

11. Is the use of gasoline for any purpose other than a fuel for internal combustion engines strictly prohibited? Are suitable solvents or waterless cleaning pastes provided where men need them, so they will not be tempted to use gasoline?

12. In cold weather, are salamanders observed at all times by a reliable fire watch?

13. After an automatic sprinkler system has been put in operation or a fire alarm system has been energized, are they checked frequently to detect tampering? Have there been any fires of suspicious origin? Have all fires been thoroughly investigated, regardless of whether serious damage to property or injury were involved?

Forms and Shoring

1. Are vertical shores truly plumb? Are they in good condition?
2. Is there any sagging or deformation of forms?
3. Is the schedule for stripping forms in accordance with accepted practice, considering weather, structural design, and any guidelines from design professionals.
4. Is the foreman in charge of stripping forms experienced in the work and aware of hazards inherent in the job?

General Observations

1. First impressions can be significant, even though the observer cannot immediately put his finger on what he senses is wrong. Are workers wearing loose or torn clothing that could catch on moving machinery or projections? Are they going about the job in a purposeful manner or is there horseplay?

2. Do craftsmen appear to be skilled? Is their work well done? Are their tools in good condition or are there split handles, loose heads on hammers, mushroomed chisels, bent screwdrivers, and folding rules that are broken off? Do you get the impression that they are likely to be concerned about safety?

3. Have OSHA notices been posted as required? Are there other safety posters, slogans, or a safety display of any kind? Do posters appear to have been there a long long time, so that nobody pays attention to them anymore?

4. Is there a paging system so supervisors or people trained in first aid can be reached quickly in an emergency?

5. Are workers talking to each other in a foreign language? If so, are bilingual safety notices and posters used? Has anyone determined whether every worker can read?

6. Are standard hand signals being used?

7. Are all accidents and near misses thoroughly investigated? Are reports brought to the attention of supervisors and workers who might be involved in similar incidents?

8. Are there frequent tool box safety meetings? Are there meetings before critical operations are begun, to be sure that all concerned are aware of the importance of each step and the hazards involved?

Guard Rails and Barriers

1. Are roads or truck routes well marked and clearly separated from the walkways that workers must use? If not, what precautions are taken to avoid accidents to inattentive pedestrians?

2. Are ramps or walkways protected with guard rails, including toe boards?

3. Are prescribed railings or wire rope barriers placed at exposed edges of floors?

4. Are open trenches and excavations protected by barriers? Are they marked with lights if there is any traffic after dark?

5. Are barriers placed to enclose the area in which workers could be struck by part of a crane as it swings?

6. Are floor openings guarded or boarded over when not in use?

7. Are public sidewalks adequately protected from falling debris? Where traffic onto the site crosses sidewalks, are there guards to direct traffic?

Hoists

1. Who is responsible for regular inspections? Are they carried out and recorded?

2. Is the enclosure adequate to prevent people from being struck by any moving part of the hoist or any materials on it?

3. Are gates provided at all landings? Are they kept shut except when the hoist is at the landing?

4. Are controls arranged so that the hoist can be operated from only one position at a time? Is there a proper signaling system?

5. Is the safe working load limit clearly marked?

6. If the hoist is for materials only, is there a notice on the cage prohibiting people from riding on it? Is the notice obeyed?

Housekeeping

1. Is refuse put into trash containers at the end of every day? Are containers emptied soon after they are full?

2. Are rags or wipers that are contaminated with flammable or toxic materials put into designated safe containers immediately after use? Are those containers emptied regularly? Are wastes disposed of in an approved manner?

3. Are all materials piled so they cannot fall or be blown about?

4. Are nails removed from scrap lumber or bent over?

5. Are all spills cleaned up promptly?

6. Is the site kept free of empty food containers, wrappers, or discarded food that could attract rats or vermin? Are there covered garbage cans for such trash? Are they used? Are they emptied regularly?

7. Are unused slings, hoses, and pallets properly stored to prevent them from obstructing traffic or becoming a tripping hazard?

8. Is glass taped or otherwise marked to make it visible to workers?

Ladders

1. Are all ladders free from split rails, loose rungs, or other obvious defects? Are ropes and metal parts of extension ladders in good condition?

2. Are ladders tied off near the top even when used for only a short time? Are workers careful to secure the bottom of the ladder against slippage whenever it is moved?

3. When ladders are used to gain access to scaffolding or floors, do they extend at least three feet above the working level? If not, are handholds provided?

4. Are damaged ladders that are to be discarded cut up immediately to prevent their being used? Are those that are to be repaired properly tagged and set aside to assure that they will not be used?

5. Are folding stepladders properly used? Do workers recognize the danger of using them in the folded position instead of a straight ladder? Do they avoid standing on the top two steps?

6. Do workers use handlines to lift tools or materials instead of carrying them up ladders in their hands?

7. Are ladders set up at the proper slope of about 1:4?

8. When a ladder is set up near a door, walkway, or anywhere it might be disturbed or interfere with traffic, is someone assigned to hold the ladder or is some other safe method used to be certain it will not be knocked down?

Lasers and X Rays

1. Are lasers and X rays used only by trained technicians? Have their qualifications been checked?

2. Are all necessary protective devices used?

3. Are unauthorized people kept out of the danger areas by posting guards or other positive means rather than merely putting up warning signs?

Machine Tools

1. Are dangerous parts such as saw blades, gears, belts, and shaft couplings adequately guarded?

2. Are saws equipped with anti-kickback pawls? Are push sticks used to avoid the need for men to get their hands near the blade? Do workers wear goggles to keep sawdust out of their eyes?

3. What precautions are used when tools create harmful dusts? Do workers recognize the dangers associated with silica and asbestos?

4. Are tool grinding wheels properly guarded? Are they kept trued up? Are tool rests within ⅛ inch of the wheels? Do workers wear goggles while grinding?

5. Are motors grounded?

6. Would it be possible for an unauthorized worker or trespasser to use the equipment? Who is responsible for securing it at the end of the day and whenever it is unattended?

Manual Handling

1. Is work planned to minimize manual lifting?

2. Do workmen know how to lift properly in order to avoid injuries? Do they recognize the dangers of improper lifting? Have there been safety meetings on the subject? Has any pre-employment screening been done to determine whether applicants have had back injuries or hernias?

3. Are rollers, conveyors, dollies, and hand trucks used whenever possible to minimize the need to carry heavy loads?

4. Are materials ever tossed from man to man? Could misses endanger others? Would it be better to use a bucket and hand line or some other method of moving the material?

5. Do workers wear safety belts with lanyards secured to a life line or substantial support whenever they must work close to an unguarded opening?

Motor Vehicles and Machines

1. Are vehicles and machines kept in good condition? Who checks them? What follow-up is done to be sure items that affect safety are corrected before work proceeds?

2. Do workers check steering, brakes, and operating controls every shift?

3. Do operators use good judgment in controlling speed, moving over rough ground, avoiding slopes that could cause rollover, and handling loads that are unstable?

4. Who checks the loading of trucks before they leave the site to be sure there will be no spillage onto public highways, and to detect the unauthorized removal of anything from the job?

5. Do men understand the dangers to which they are exposed when working behind dump trucks, unloading pipe or piling, releasing part of a load of trusses that have been tied together during transit, or performing similar tasks? Does anyone check to be sure each worker has been trained before he is assigned a new task that could be dangerous?

6. Have bridges over which heavy trucks must pass been inspected and approved for the weights to which they may be called on to carry? Would it be dangerous for more than one truck to be on the bridge at once? What speed is permissible?

7. Are any oversize loads to be moved to the site? If so, are permits required? Do they restrict the route, the time of day, or day of the week when the load may be moved?

8. Who may ride the machine or truck? Are rules against unauthorized riders enforced?

9. Who maintains pneumatic tires on trucks and earthmoving equipment? Are the workers well trained in safety and equipped with tire cages and other special tools to prevent injury in case something goes wrong while a large tire is being inflated.

10. Is every motorized tool or motor vehicle parked in a safe position and securely locked whenever its assigned operator is absent from the immediate vicinity?

11. Are overhead guards, rollover protection, and safety belts installed when required and consistently used?

12. When backing, what precautions are taken to prevent injury to people or damage to property in the operator's "blind spots"? Are warning bells needed?

13. Is the driver required to leave the cab while his truck is being loaded or unloaded by a crane whenever collapse of the crane could put him in danger?

14. Do operators and maintenance men recognize the danger that fluid may penetrate the skin and cause very serious injury if there is a leak in a high pressure hydraulic system?

15. Are blocks placed to support the blade or any moving part before anyone reaches under it?

Noise

1. Are there any very noisy machines or tools? Do any men have to shout to make themselves understood by a person a few feet away? If so, have noise level tests been made to determine whether it is permissible for a worker to be exposed to the measured noise for the full workday?

2. If the noise is too loud (over 90 dBA), can it be reduced by installing mufflers, repairing the equipment, or enclosing it? Would it be feasible to increase the distance from the source of noise to the workers?

3. Should ear plugs or muffs be worn?

4. Are any sudden noises of over 140 dB produced by any tools or equipment?

Powder-Actuated Tools

1. Has the tool operator been properly trained? What certification has he to prove that he is qualified to use the make and model tool required for the job?

2. How has it been determined that the tool is appropriate and safe? Is there any possibility that a fastener might be driven entirely through a wall and become a projectile?

3. Does the worker wear goggles? Does he know how to deal with misfires? Does he keep the tool clean? Does he follow the manufacturer's safety instructions?

4. Are cartridges kept in a secure place? Is the tool safe from unauthorized use when the assigned operator is not there?

Protective Clothing and Respirators

1. Are hard hats required? If so, have suitable ones been obtained? Are they being worn?

2. What type of gloves should be used for each operation? Are men using suitable ones? Are they in good condition?

3. What type of respirator is needed for each type of work? Is it being used? Is it kept in sanitary condition and in good working order? What air quality tests are made at critical areas?

4. Do workers have suitable footwear? Should they be wearing safety shoes for toe protection, guards over their shoes, insulating soles, spark-proof shoes, or cold weather boots? Are soles badly worn, slippery, or unsuitable for the job?

5. Do workers have and use goggles when doing work that is dusty or throws off sharp particles? Do they wear colored goggles of the proper shade to protect their eyes from welding glare? Are men wearing bifocals prohibited from working at elevations or where they might trip if they misjudged the distance? Should a face shield with a support for prescription lenses be used instead of goggles and a respirator?

6. Should any workers have special protective gear to prevent injury by splashes of acid, molten roofing materials, or strong caustics?

7. Are regulations regarding the handling of asbestos being observed?

8. Are safety belts, harnesses, and life lines provided and used where needed? Are workers aware of the danger of impalement when working even a few feet above projecting reenforcing rods?

Roofing

1. On sloping roofs, are crawling boards, life lines, and edge protection provided where needed?

2. Are regulations governing work on flat roofs being obeyed?

3. Are there weak spots, skylights, or deteriorated asbestos-cement boards through which a worker might fall?

4. Are safe methods being used to melt and handle bituminous materials?

5. Are machine tools such as cutters and power brooms raised onto flat roofs and removed from them in a safe manner? Do workers stay out from under while these moves are being made?

6. Are foremen alert to the possibility that a worker may be reaching his physical tolerance for heat or exertion? Do they anticipate problems rather than wait for trouble?

Sanitation and First Aid

1. Are approved toilet facilities available? Are washbasins and clean water available? Are workers required to wash after using chemicals or after exposure to sewage or other possibly contaminated materials?

2. Is there a suitable place for workers to store food and to eat?

3. Is there an adequate supply of safe drinking water? Is it kept safe from contamination on the job? Is there a bubbler or are sanitary cups available?

4. Has a first aid kit been obtained and approved by a doctor or other qualified person? Is it kept supplied? Is someone always available to render first aid? Do workers know who he is?

5. Have harmful materials been identified? Have the necessary precautions for their safe use been determined? Are they being followed?

6. Is safety gear regularly checked, cleaned, and replaced if defective?

7. Are danger areas restricted to workers who must enter them? How are others excluded?

8. Is a permit system in use to control work in confined spaces? Are all necessary steps taken to prevent accidents? Are permits valid only for specific and limited times?

9. Is welding or burning of lead-painted metal, galvanized steel, cadmium-plated steel, or any other material that could give off toxic fumes prohibited without specific approval of the safety engineer?

10. Are workers prohibited from smoking after handling hazardous or contaminated material until they have washed thoroughly. Do they know how to loosen and shake off gloves without getting irritants on their hands?

Scaffolds

1. Are all scaffolds designed and erected by competent persons? When the regulations require that the services of a professional engineer be provided, is this done?

2. How was the required loading limit determined? Are loads distributed as uniformly as contemplated by the designer? If work is interrupted by masons, when do other workers stop bringing more material up to the working level? Who calls a halt when the load limit of the scaffold has been reached?

3. Are scaffolds inspected every day before work begins? Does the inspector check to be sure no braces or other essential parts have been disturbed?

4. Are all uprights provided with base plates or footings to assure proper support and stability, even in wet weather? Does water collect around the base plates and soften the ground?

5. Has proper access been provided to the working levels? Are ladders securely clamped or lashed in place? Do they extend at least three feet above the working level, or are handholds provided?

6. Are planks in good condition? Are they sufficient to cover the whole width of the work area, out to the toe board, without leaving space for material to fall or for a worker's foot to pass? Are boards properly supported to prevent displacement?

7. Are railings, including midrails and toe boards, provided in accordance with OSHA rules?

8. Is the scaffold secured to the building at enough points to assure stability?

Security

1. When any unidentified person walks onto the site, is he required to wait until the person he wants to see arranges to meet him?

2. How do drivers making deliveries find the people who are to sign for them? Could anyone pose as a driver and move about the job without being challenged?

3. Is the site enclosed? Is the fence inspected before work begins and at the end of the day? Are gates locked? Is there a watchman or security system? Would it be possible for anyone to remain after work without being detected?

4. Could vandals, children, or inquisitive people get onto the site? Could they get onto scaffolds or tamper with machines?

5. Are petty thefts investigated? Have there been any serious incidents of vandalism, disappearance of materials, or fires of suspicious origin?

6. Is a log kept of all visitors, showing name, business, who they saw, when they arrived, and when they left? Are badges given to visitors, and returned when they leave?

7. Have there been any demonstrations or threats of violence? Have plans been made to cope with attempts to stop work?

Weather

1. If there is an exceptionally severe rainstorm, is flooding likely? Will additional shoring of excavations be needed if rain continues? Will pumps be needed? Is emergency power available to operate them?

2. Is the structure adequately braced or guyed to withstand winds that have been forecast, with due allowance for sudden gusts in excess of predictions? Are there any loose materials that should be secured?

3. Are cranes left in a safe position? Are machines parked where they would be safe in case of a mud slide or similar shifting of the earth?

4. Is shelter provided for workers in case of rain, snow, or icing con-

ditions? Who determines when work must be stopped because weather changes have made it unsafe to continue?

5. Are access roads, walkways, and work platforms free of snow or ice and safe to use?

6. In hot weather, are first aid people prepared to treat cases of heat exhaustion and sunstroke? Are preventive measures taken in anticipation of these illnesses?

APPENDIX L
Biography of Author

James B. Fullman earned his Bachelor of Engineering degree at Yale University in 1936. He was a pioneer in radiant panel heating and snow melting and during WW II he was in the Navy with responsibility for installation and repair of degaussing gear on merchant ships in the New York area. He was Chief Engineer of Water Service Laboratories when that firm was acquired by Olin Corporation. When OSHA emerged, he prepared a compliance manual and was actively involved in safety problems until his retirement in 1976. At that time he was Manager of Manufacturing and Distribution Projects, Olin Water Services.

Mr. Fullman is a Professional Engineer; Accredited Corrosion Specialist, National Association of Corrosion Engineers; holder of two patents; and co-author of *Water and Piping Problems in Large and Small Buildings, A Troubleshooters Guide,* published by John Wiley and Sons in 1981.

Index

Abrasive wheels, 200, 206
Absenteeism, 12
Acetone, 177
Acetylene *vs.* 176-177
Acids, 244, 250
Adaptation, 20, 47
Air, 211. *See also* Confined spaces
Airconditioning, 171
Alcoholism, 5, 26
Alcohols, as solvents, 249
Allergies, 225
Aluminum, 175, 178
Ambulance service, 68
American Institute of Steel Construction, 190
American Insurance Association, 44
American Mutual Insurance Alliance, 44
American National Standards Institute, 83, 142, 206, 254
American Society of Civil Engineers, 19, 183
American Water Works Association, 125
Anger, 13
ANSI, *see* American National Standards Institute
Anxiety, 12
Arbitration, 195
Archaeology, 98
Arson, 73, 177
Asbestos, 43, 101, 107, 239, 247
Association of American Railroads, 42
Attitudes, 3-13, 195
  early development of, 197
  toward OSHA, 156
  of owners, 135
  toward use of ladders, 167
  toward use of safety belts, 158-160
Attractive nuisance, 99
Audio-visual training, 97
Authority, lines of, 51, 111
Automobiles, 138
Awards, 237
Awareness, 21, 31
Axes, 105

Back filling, 123
Back injuries, 115, 145
Bacteria, 86
Barricades, 122, 142, 270
Batteries, 128
Bechtel Corporation, 14, 49, 100, 168
Bees, 106
Belts, safety, 153, 158-161, 225
Bench grinders, 206
Benzene, 247
Bids, competitive, 186, 227
Bifocals, 138
Biocides, 235
Biography, author, 279
Black Lung Act, 60
Blasting, 35, 103, 266
Body harness, 161
Bonding, electrical, 174
Bosun's chair, 161
Breathing apparatus, self-contained, 219
Bribes, 51
Bridges, 157, 177, 186, 197
Brooklyn Bridge, 186
Building trades unions, 42-44
Burns, 136, 179, 224, 250
Business Periodicals Index, 45

Cadmium, 213
Cans, safety, 117
Carbon dioxide, 213, 247
Carbon monoxide, 213, 248
Carbon tetrachloride, 174, 248
Cardiopulmonary resuscitation (CPR), 130
Cartridges, for respirators, 88
Causes of accidents, 2, 8-13, 144, 183, 195
Caustics, 223, 251
Ceremonies, 237
CFR, *see* Code of Federal Regulations
Chain, 143
Chain saws, 105

Check list:
  insurance, 55
  safety hazards, 266-278
Chemical(s), 223
  cleaning solutions, 107
  dumping of, 98
  feed pumps, 222
  hazards, 181
  wastes, 83, 102
Cherry pickers, 141
Chicken ladders, 182
Chisels, 201
Chlorine, 223
Cleaning solutions, 222. *See also* Chemicals
Clearing of sites, 104
Cocoa Beach, Florida, 184
Code, National Electric, 129
Code of Federal Regulations, 31. *See also specific hazards*
Combustible liquids, 173
Community relations, 79, 99
Compressed air, 117, 221
Compressed Gas Association, 32
Concrete:
  failure of, 154
  placing of, 130, 154
  strength of, 183
  welding on, 176
Confined spaces, 211-219
  classes of, 217
  entry permit, 264
Construction Industry Manufacturer's association, 142
Consultants, safety, 39
Containers, empty, 106
Contaminants, air, 215
Continuity of safety programs, 48
Contributions, 51
Conveyors, 145
Cooperative safety programs, 39
Coordinator, designated safety, 17
Copper *vs.* acetylene, 176
Corrosion, 126
Corruption, 51, 227
Cost effectiveness, 3-6, 18, 62, 167, 228
Costs of accidents, 6-8, 58, 155, 179
CPR, *see* Cardiopulmonary resuscitation
Cranes, 116, 130, 138-144
  check list, 266
  collapse of, 10, 16
  contact with electric power lines, 140
  handling piling with, 116
  licensing operators of, 138
  placing concrete with, 138
Cranking of engines, 117
Crawling boards, 182
Cross connections, 125, 220, 236
Cutting and patching, 206

Dampers, 171
Dangerous substances, 246-251
Decibels, 71
Demolition, 100, 239
Demonstrations (protests), 81
Department of Transportation, *see* Transportation, Department of
Depression, 12
Derricks, 142
Disciplinary action, 37-38, 51
Diseases, 138
D.O.T., *see* Transportation, Department of
Drawings, as built, 233
Drills, electric, 202
Drowning, 116
Drugs, 12, 69, 76, 139, 209
Dunnage, 136
Du Pont (E. I. du Pont de Nemours and Company), 19
Dust, 86, 117, 226
Dyslexia, 95

Ear protection, 73
Earth moving equipment, 113
Earthquakes, 77, 183. *See also* Insurance
Electric power, 127, 140, 155, 220, 267
Electrocution, 9, 128, 140, 170
Emergencies, anticipation of, 66, 134
Empty containers, 106
Enclosed spaces, 211-219, 264
Entry permit for confined spaces, 264
Environmentalists, 99
Environmental Protection Agency, 107, 124, 223
EPA, *see* Environmental Protection Agency
Epoxies, 212, 225
Equipment, safety, for operating engineers, 231
Ethylene chlorohydrin, 249
Excavations, 110, 118, 268
Explosives, 35, 103, 106, 203
Explosive vapors, gases, and mixtures, 9, 107, 176, 215
Extension cords, 253
Extinguishers, fire, 179
Exxon Corporation, 15
Eye protection, 84, 87, 223

Facades, 193

Face shields, 84
Falls, 9, 155-171, 179, 212, 224
Family influence, 48
Fatality rates, 1, 4
Fatigue, 11
Fear, 27
Federal Depository Libraries, 41
Federal Regulations, Code of, 31
Fights, 13
Fire, 9, 105, 123, 125, 171-179
  alarms, 224
  check list, 268
  classes of, 178
  extinguishers, 142, 179
  flashover, 178
  insurance, 55
  protection, 73
  resistance, 171
  vs. steel, 178
First aid, 66, 130, 134, 275
Flammable liquids, 107, 173
Flashover, 178. *See also* fire
Flash point, 173
Fleas, 106
Float valves, 219
Floors, 157
Fluorescent light tubes, 107
Fogging of goggles, 84
Food, 208
Footwear, 92
Forms, 131, 154, 269
Frequency of accidents, *see* Incidence rates
Fumes, 86

Gases, 85
Gas free engineer, 217
Gasoline, 117, 174, 176, 181, 249
Gauntlets, 91
Geological formations, 119
Gifts, 51
Glare, 123
Glass, 193
Glomar Pacific, 15
Gloves, 91
Goggles, 84
Gratuities, 51
Grounding connections, 129
Guard railings, 157, 162, 182, 259-260, 270. *See also* Falls

Hammer tests, 221
Hand injuries, 91, 145, 224
Hand railings, *see* Guard railings
Hand tools, 197-206
Hard hats, 83, 102, 129
Hartford Coliseum, 189, 197, 232
Health and Safety Executive (British), 46
Hearing protection, 70
Heart disease, 9, 11, 138
Heat exhaustion, 11
Hernia, 115, 145
Hertz (Hz), 71
Historical societies, 99
Hoists, 142, 270
Hornets, 106
Hoses, 117, 125
Hospitals, 67
Housekeeping, 173, 207, 271. *See also* Fire; Trash disposal
HSE, *see* Health and Safety Executive
Hyatt Regency Hotel, 187
Hydraulic pressure systems, 115, 224
Hydrochloric acid, 249
Hydrogen, 128
Hydrogen sulfide, 126, 248
Hz (Hertz), 71

Illinois, University of, 183
Illiteracy, 94
Illness, 11, 68
Illumination requirements, 254
Impalement, 133
Incentives, 22, 28, 118
Incidence rates, 1-3, 13, 24, 242
Independent contractors, 59
Indirect costs of accidents, 6
Injuries:
  medical services required for, 66
  types of, 67
Insects, 106, 208
Inspections, 142, 227
Instructions, 113, 197, 232, 230-232
Instruments, 215
Insulation, 172
Insurance, 6, 44, 53-66, 233
  check list, 55
  claims, 145
  uninsured contractors, 55
  workers compensation, 6, 56-66
Interferences, 206

J. C. Penny Company, Inc., 192
Jurisdictional disputes, 194

Kemper Arena, 189
Keys, 76, 209, 234
Knots, 148-153
Knowledge of hazards, 20, 31-47

Labor, Department of, 1
Laboratories, 223
Labor conflicts, 194
Labor Statistics, Bureau of, 245
Ladders, 145, 156, 166-170
  check list, 271
  cleat type, 261
  for electrical work, 129, 170
Lasers, 109, 123, 272
Lead, 224
Leadership, 21
Liability insurance, *see* Insurance
Libraries, Federal Depository, 41
Licensing, 17, 138
Lifelines, 153, 182
Lifting, manual, 145
Lighting, 254
Litigation, 195
Local regulations, 45, 50
Lock-out system, 220, 234
Longshoremen's and Harbor Worker's Compensation Act, 57
Lost workdays, 242

M (as an abbreviation), 223
Machine tools, 272
Management, 18, 23
Manual lifting, 145, 196, 272
Manufacturers, 46
Manufacturing, 2, 14
Marital difficulties, 12
Marquette University, 183
Materials:
  handling, 136, 144, 179
  selection of, 172
Medical:
  examinations, 138, 145
  services, 66
Medication, 69
Meetings, 167
Mental health, 12
Methane, 213
Methanol, 249
Methyl chloroform, 248
Metric system, 42, 144
MGM Grant Hotel, 171
Midrails, 162
Mine Safety Appliances Company, 214-216
Mining, 2, 14
Minorities, 25
Mists, 86
Monitoring, 20, 182
Motion stopping systems, 182
Motivation, 20-31. *See also* Attitudes
Motors, 201, 220, 222
Motor vehicles, 273
MSA, *see* Mine Safety Appliances Company
Muggings, 211
Muriatic acid, 249

Naptha, 174
National Electric Code, 129
National Fire Protection Association, 44, 130, 173
National Institute for Occupational Safety and Health, 40, 82, 88, 179, 203, 211
National Safety Council, 42, 145
National Technical Information Service, 41
Near misses, 30, 47, 135, 243
Negligence, 7. *See also* Insurance
Nets, safety, 157
News coverage, 79-82
NFPA, *see* National Fire Protection Association
Nibbler, 202
NIOSH, *see* National Institute for Occupational Safety and Health
Nitrogen, 211, 247
Noise, 70-73, 117, 274
NTIS, *see* National Technical Information Service

Occupational Safety and Health Act of 1970, 33
Occupational Safety and Health Administration, 5, 15, 33-44, 157
  regional offices of, 244
Odors, 87
Olin Corporation, 204
Organization, 19, 183-193
OSHA, *see* Occupational Safety and Health Administration
Over-regulation, 163, 165, 173, 182, 246
Owners, 18, 135, 183-193, 227
Oxygen, 176, 215, 221
  deficiency of, 87, 211-219, 255, 264

Pails, 146, 151
Painting, 43, 212, 224
Pallets, 136, 235
Panama Canal, 121
Parking of equipment, 114, 141
Penalties, 33, 187
Pentachlorophenate, 251
Personal safety equipment, 82, 274. *See also specific entries,* eg., Shoes, Gloves, Goggles
Philadelphia Electric Company, 169

Piles, 110, 116
Pipe, 122, 137, 145
  sleeves, 206
Pitch, 181, 247. *See also* Burns
Planks for scaffolding, 165
Planning of safety programs, 18-49, 134
Plan review, 183-193
Plumbers, 155
Poison ivy, 105
Policy statements, 23
Polyvinyl chloride, 175
Posters, OSHA, 35, 47, 142, 240
Powder actuated tools, 203, 274
Power tools, 199
Pressure relief valves, 221
Pressure testing, 176, 221
Prizes, 48
Programs, safety, 18-49, 134
Property damage insurance, 54
Protective clothing, *see* Personal safety equipment
Psychological problems, 12
Publicity, 15, 79, 135, 237
Public safety, 185, 227
Purchase orders, 136
Purdue University, 183
PVC, *see* Polyvinyl chloride

Quick fill lines, 219

Radio broadcasting, 141, 142
Railings, *see* Guard railings
Receiving, 137, 209
Recordable injuries and illnesses, 239-241
Recordkeeping, OSHA, 239
Records, *see* Incidence rates
Reenforcing rods, 133
Refrigerants, 248
Regulations:
  effectiveness of, 15-17
  enforcement of, 36-40
  local, 45, 50
  source of, 195
  see also Over-regulation
Relief valves, 118
Repairs, machine, 115
Reprimands, 70. *See also* Disciplinary action
Reports, inspection, 232
  accident, 6, 240-243. *See also* Near misses
Respirators, 85, 117, 218, 255, 274
Reviews, safety programs, 134, 237
Rigging, 143, 196
Rings, 225. *See also* Hand tools
Robbery, 75
Rocky Mountain Spotted fever, 106
Rodents, 208
Role playing, 21
Rollover protection, 114
Roofing, 5, 155, 179-182, 275
Rope, 143-147. *See also* Knots

Sabotage, 74
Safety equipment, *see* Personal safety equipment; *specific entires,* e.g., Gloves
Safety factors, 139
Sanitation, 275
Saws, 200
Scaffolding, 43, 156, 163-166, 258-260, 276
Scavengers, 99
Screw drivers, 198
Security, 76, 205, 209, 233, 277
  systems, 224. *See also* Keys
Seminars, 21
Service agreements, 234-237
Service lines, 108
Sheet metal workers, 155
Shoes, 92, 103
Shoring of excavations, 118-122, 268. *See also* Blasting
Signals, 142
Signs, 49, 100, 142
  bilingual, 95
Silicosis, 107
Sink holes, 98
Site inspections, 98
Size of company, 14, 243
Skil Corporation, 199, 253
Slings, 143
Slippage of knots, 148, 153
Sloping of excavations, 119
Smoke alarms, 172
Smoking, 173, 268
Snakes, 106
Sodium azide, 102
Sodium silicate, 251
Soil types, 119
Solvents, 178, 224
Spiders, 106
Spontaneous combustion, 175
Spores, 86
Sprinkler systems, 172
Squatters, 98
Staking out, 108
Standards, 15, 35. *See also* American National Standards Institute
Stanford University, 6, 155
Static electricity, 174, 249

Steam, 126, 130
Steel erection, 135, 155-157, 183-184
Steel *vs.* fire, 178
Sterilizing, 223
Stouffer's Inn fire, 232
Stress, emotional, 12
Structural collapse, 108, 178, 183-192
Structural engineering, 183, 185
Subrogation, 53
Suicide, 12, 28
Sulfuric acid, 250
Sun stroke, 11
Superintendent of Documents, 41
Surveying, 108

Tardiness, 12
Temporary power, 127
Terminology, 31, 96
  of air contaminants, 86
  chemical, 223
  fire, 173
Testing, 219
Theft, 75, 170, 207, 209. *See also* Security
Threshold Limit Value (TLV), 87
Ticks, 106
Tire cages, 115
TLV, *see* Threshold Limit Value
Toe boards, 162
Tools, small, 197
Torch burning, 175
Traffic, on site, 112
Training, 19, 35, 43, 94-97, 211
  limitations, 22, 95
  operating engineers, 230-232
  physical, 131
  roofers, 180
  safety engineers, 21, 96
Tranquilizers, 69
Transformers, 128, 220
Transportation, Department of, 41, 103, 107, 137, 173. *See also* Bridges
Trash disposal, 102, 106-108. *See also* Housekeeping
Trisodium phosphate, 251
Trucks, 137
Tube and coupler scaffolds, 260

Underground obstructions, 108
Unions, 42, 51, 70
United States Department of Labor, *see* Labor, Department of
United States Department of Transportation, *see* Transportation, Department of
United States Steel Corporation, 184
Unloading, 137, 144
Utilities, 110

Vacuum breakers, 125
Vandalism, 75
Vapors, 86
Vehicles, 138
Vibration, 117. *See also* Noise
Vibration white finger disease, 92, 202
Vinyl chloride, 247, 248
Violence, 25

Wall openings, 163
Wasps, 106
Waste treatment, 246
Water:
  potable, 124
  supplies, 123-126
  treatment, 125, 235
Wausau Insurance Companies, 44, 179
Weather, 11, 180, 277
Welding, 85, 123, 175-177, 221, 252
Wells, 124. *See also* Water supplies
Wheelbarrows, 145
Wheel chocks, 131
Wind, 11, 107, 144, 183. *See also* Weather
Wire rope, 142-144
Wood:
  fire resistance of, 178
  groups of, 262
  preservatives for, 251
Worker's compensation insurance, 6, 55-56
Wrenches, 199

X-rays, 272